John Charles Van Dyke

Nature for its own Sake; first Studies in natural Appearances

John Charles Van Dyke

Nature for its own Sake; first Studies in natural Appearances

ISBN/EAN: 9783337024635

Printed in Europe, USA, Canada, Australia, Japan

Cover: Foto ©berggeist007 / pixelio.de

More available books at **www.hansebooks.com**

BOOKS BY PROF. JOHN C. VAN DYKE

Art for Art's Sake. University Lectures on the Technical Beauties of Painting. With 24 Illustrations. 12mo.

The Meaning of Pictures. University Lectures at the Metropolitan Museum, New York. With 31 Illustrations. 12mo

Studies in Pictures. An Introduction to the Famous Galleries. With 40 Illustrations. 12mo

What is Art? Studies in the Technique and Criticism of Painting. 12mo

Text Book of the History of Painting. With 110 Illustrations. New Edition. 12mo

Old Dutch and Flemish Masters. With Timothy Cole's Wood-Engravings. Superroyal 8vo

Old English Masters. With Timothy Cole's Wood-Engravings. Superroyal 8vo

Modern French Masters. Written by American Artists and Edited by Prof. Van Dyke. With 66 Full-page Illustrations. Superroyal 8vo

New Guides to Old Masters. Critical Notes on the European Galleries, Arranged in Catalogue Order. Frontispieces. 12 volumes. Sold separately

American Painting and Its Tradition. As represented by Inness, Wyant, Martin, Homer, La Farge, Whistler, Chase, Alexander, Sargent. With 24 Illustrations. 12mo

Nature for Its Own Sake. First Studies in Natural Appearances. With Portrait. 12mo

The Desert. Further Studies in Natural Appearances. With Frontispiece. 12mo
Illustrated Edition with Photographs by J. Smeaton Chase.

The Opal Sea. Continued Studies in Impressions and Appearances. With Frontispiece. 12mo

The Mountain. Renewed Studies in Impressions and Appearances. With Frontispiece. 12mo

The Grand Canyon of the Colorado. Recurrent Studies in Impressions and Appearances. With 34 Illustrations. 12mo

The Money God. Chapters of Heresy and Dissent Concerning Business Methods and Mercenary Ideals in American Life. 12mo

The New New York. A Commentary on the Place and the People. With 125 Illustrations by Joseph Pennell.

NATURE
FOR ITS OWN SAKE

FIRST STUDIES IN NATURAL APPEARANCES

BY

JOHN C. VAN DYKE

AUTHOR OF "ART FOR ART'S SAKE"

NEW YORK
CHARLES SCRIBNER'S SONS
1921

PREFACE

THE title and the treatment of this book require a few sentences of explanation. The word "Nature," as it is used in these pages, does not comprehend animal life in any form whatever. It is applied only to lights, skies, clouds, waters, lands, foliage—the great elements that reveal form and color in landscape, the component parts of the earth-beauty about us. In treating of this nature I have not considered it as the classic or romantic background of human story, nor regarded man as an essential factor in it. Nature is neither classic nor romantic; it is simply—nature. Nor is it, as some would have us think, a sympathetic friend of mankind endowed with semi-human emotions. Mountains do not "frown," trees do not "weep," nor do skies "smile"; they are quite incapable of doing so. Indeed, so far as any sympathy with humanity is concerned, "the last of thy brothers might vanish from the face of the earth, and not

a needle of the pine branches would tremble."
"Nature for its Own Sake," then, means simply that herein nature is considered as sufficient unto itself. The forms and colors of this earth need no association with mankind to make them beautiful.

So far as application or illustration is concerned, my argument has no direct bearing upon any branch of science, literature, or art. I have used scientific facts occasionally to point a meaning without designing a scientific book; I have in places spoken of literature, but the book is not an appeal to nature from those who have written about it; and as for art, the word does not appear after this preface. Painters or writers, with their truth or falsity of statement, are not my present concern. What, then, is the object of the book? Simply to call attention to that nature around us which only too many people look at every day and yet never see, to show that light, form, and color are beautiful regardless of human meaning or use, to suggest what pleasure and profit may be derived from the study of that natural beauty which is everyone's untaxed heritage, and which may be had for the lifting of one's eyes.

In measure these pages are records of per

sonal impression, and must be so regarded by the reader. However objective in treatment one might wish to be, his point of view is always more or less warped by the personal equation, and I can pretend to nothing more than a—view. As the sub-title indicates, these impressions are general in character—in fact "first studies." The book is designed as an introduction to a subject which I hope to consider more fully hereafter.

SAGE LIBRARY,　　　　　　　　　　J. O. V. D.
NEW BRUNSWICK, N. J.

CONTENTS

CHAPTER I. *Pure and Reflected Light.*—Our knowledge of light—How light is obstructed by atmosphere—Dust and vapor particles in the air—Pure sunlight violet-blue—White light the residue after filtering through the air—Differences in light—Shown in cleared air—And from mountain-tops—Earth coloring regulated by density of the air—Warm and cold colorings—The dust veil—Krakatoa and the red skies—Color in the tropics—And at the poles—Sunlight in summer—The dawn—In Egypt with its wings of light—In temperate climes—Flooding of the dawn-light—The sunrise—Sunrise colors—Noon-light—Its great beauty in the fields—The fall of light—Sunset—Red sun disks—Sunset colors—Spectrum colors on the sky—Sky effects—Twilight—Zodiacal light—The moon and its rise—The misshapen moon—Twilight and moonlight blended—Horizon hues at twilight—Star-light—Star-colors—Darkness of the upper space—Lights of the night—The great variety—Countless changes—Beauty of the world 1

CHAPTER II. *Broken and Shaded Light.*—Cloud light and the cloud veil—The lowery day—Rain clouds—Storm light—Night and storm clouds—Mists and fogs—Vapor lights—White horizons—Fog lights—Fog and smoke—Fog effects—Color in fog banks—Nature's delicate hues—Alternate sunlight and cloud light—Sun-bursts—The fall of sunbeams—Sun-shafts with rain—The sun "drawing water "—Sun-bursts and flying shadows in Scotland—Moon-bursts and moon beams—Shaded light—The law of shadows—Electric-light shadows—Shadows lightened by light-diffusion—Shadows in hot weather—The colored shadow—Scientifically explained—Complementary hues in shadow—Necessary conditions of the colored shadow—Blue shadows on snow—Lilac shadows on clay and sand—The mixed colors in nature—Shadow complications—The shadowless day—Odd colors in shadow—Shadows of the moon—And of the stars...................... 25

CHAPTER III. *The Blue Sky.*—Impressions of the sky—Transparency of the blue—Sky depth—Through the clouds—Sky reach—Sky lines seen at sea—Horizon lines—Sky lines seen from heights—Apparent changes across the face of the sky—Sky waves and undulations—The blue seen from mountain-tops—The Great Silence of the firmament—The blue seen from the valleys—Its changes by day and by night—The tenderness of its coloring—Our non-observance of the sky—And of sky tints—Alpine glows at home—Skies in different lands—Color changes through atmosphere—Season changes in the blue—Luminosity of the blue—Transmitted and reflected light—Sky lights on the earth—Reflection from the blue—Atmospheric reflection—The dawn an illustration of atmospheric reflection—And also a symbol in natural religion 47

CHAPTER IV. *Clouds and Cloud Forms.*—Cloud-making—Cloud forms—Why clouds float—Effect of the winds—Effect of the air-currents—Cloud caps—Banner clouds—Self-renewal of clouds—Clouds, how acted upon—How moved—Day and night clouds—Classification of the families—The cirrus—Whiteness of the cirrus—Color of the cirrus—The cirro-stratus—Sun and moon halos—The cirro-cumulus—Dappled and mackerel skies—The stratus—The strato-cumulus—The cumulus—Cumulus changes—Summer clouds—Heap clouds—Cloud illusions—The cumulo-nimbus—Silver linings—The nimbus—Forms of the rain cloud—Storm clouds—Scattering forms of cloud—Scud, wrack, etc.—The lightness and drift of clouds—Cloud fancies—Cloud splendor and coloring—Seen at sunset—Seen at dawn and at noon-time—Value of clouds in landscape—Seen with a low sky line.... 65

CHAPTER V. *Rain and Snow.*—The vapor-carrying capacity of air—Condensation—Causes of clouds and rain—Eastern storms, how produced—Warm winds and cold mountains—In the high Alps—The rain-drop—Size of the drop—The first heavy fall—Thunder-storms—Lightning and clouds at night—Rain fringes—Surrounded by rain—The rainbow—Three-day storms—Rainy days—After the storm—City *vs.* country rain—Hail—Its formation—The hail theories—The falling stones—Snow-flakes—Snow on the mountains and rain in the valley—The first fall—Snow-storms—The blizzard—Flying snow—The luminosity of snow—Snow prisms—Brilliancy of snow reflection—The snowy landscape—Under moonlight—Snow lines—Snow colors and shadows—Swirls and drifts—In early spring—Nature's skeleton—Nature's awakening.................................. 88

CHAPTER VI. *The Open Sea.*—First impressions—Sea-changes—Water forms—The strife of the sea—Its restlessness—Wind and wave—Wave crests—Storm waves—The hurricane sea—The height of waves—Thickness of waves—Tropical swells—Lines of a wave—Northern and Southern waves—The undulation and wave motion—Depth of the undulation—Local hues of water—Sea-floors and their influence on coloring—Deep-sea color—Gulf and bay colorings—Mineral hues in water—Color patches—Sea sawdust—Transparency of sea-color—Reflection from surface—The smooth swell on the Southern seas—Northern waters—Sky effects at sea—Sunlight on the waves—Moonlight on the waves—Cloud shadows upon water—Colored shadows again—Cloudy days at sea—The emerald-greens of storm—Atlantic and tropical waves—Following the equator............. 113

CHAPTER VII. *Along Shore.*—On the beach—The coast-wave—Why waves break—Dancing jets under a cliff—The size of coast-waves—And their power—Forced and wedged waves—The beach-comber—Water-mirrors on the beach—The undulation again—The rising of the sea—Thrust of the waves—Curves of sand beaches—Wave action on the rocks—Cliff undermining—Rock forms made by water—Pulpits, bridges, and caverns—Formation of sand-dunes—Sea barriers—Bars, lagoons, and marshes—The tides—Ebb and flood tides—The bare shores—Coast lines—Color and light upon the shore—Twilight colorings—Moonlight on the sea—The coast in storm—The whipped waves—The uses of storm—Without the sea.................................... 134

CHAPTER VIII. *Running Waters.*—The river at the sea—Meeting the ocean—The river's path—The Plain Track—Through the meadows—The river's basin—The sluggish flow—The Valley Track—The river island—Hurrying waters—New movement—The wear of water—The sculptor of the land—Valley and mountain carvings—Oscillations of the stream—Lines of the banks and the water—Color on the river—With snow and under ice—

Freshets—Floods—The Mississippi—The river as it was and as it is—European rivers—The Thames, the Rhine, the Danube—The Mountain Track—Brooks—The mountain-brook and its motion—In the ravine—The gorge—Following the brook—By the waterfall—The cataract—Niagara—Brook reflections—The frozen stream—Purity of brook waters—The river's source—Catch-basins—The rivulet—The beginning of the stream............. 153

CHAPTER IX. *Still Waters.*—Names of seas and lakes—Definitions—Lakes vs. oceans—The mountain-lake—Its various features—Purity and clarity of its waters—Lake charm and sentiment—Local coloring of the water—Colors of background—Local hue and reflection—Confusion of hues—Reflections—Seen at night—Confusion of reflections with shadows—Surface appearances and phases of reflection—On darkened waters—On strong-hued waters—Variations and distortions—The likeness inexact—The angle of reflection—Elongated reflections—The Angels' Pathway—Romance—Moonlight on the lake—Material beauty of American lakes—Lake George a type—The pond in the forest—The prairie pond—In Indian days—Artificial waters—Venetian lagoons and canals—Holland canals—The mountain-lake once more—Its serene beauty............................. 174

CHAPTER X. *The Earth Frame.*—Earth and sea—The earth's surface—Inequalities of the surface—The skeleton of the earth—Strength of the frame—Formation of the crust—Geological formations—Solidity of the earth—Permanence of the flat prairies—And of the primeval forests—And of the desert—The sands of Sahara—The vaulting of the globe—The understructure of the Alps—The base of the Jungfrau from Mürren—Foundations of mountains—The hardness of rocks—Nature's building principle—The self-supporting globe—The lines of the earth—Shadow of the earth upon the sky—The arch of the sky—Horizon lines at sea and on the prairie—The curved line and "the line of beauty"—The law of the circle—Shown in the forms of nature—And in the elements and the solar system—Circles in physical and intellectual life—The uttermost rim of thought—The vanity of progress—The universal law............. 197

CHAPTER XI. *Mountains and Hills.*—Mountain ridges —How the mountains are formed—The wrinkle or fold theory—The Alps—The age of mountains—Denudation and erosion—The old Appalachians—The worn-down mountains—Exposed crusts—Mountains cut out by water —The approach to the mountains from the plains—Seen from a distance—Mountain-climbing—The view—The panoramic scene—From the high Alps—The look downward—Distorted light and color—The look upward—The clouds and the sky—The mountains from the valley— Mountain colors—The lower ranges—Sky lines—Mountains at sunrise—At noon—At sunset—The western barrier—Looking eastward at sunset—Mountain glow at sunset—The Alps in storm—Storm in the lake-reflection —Mountain individuality—Changes of form—Of color— Influence of atmosphere—Light changes—The green hills—English hills—New England ranges—Hills in landscape—The levelling down.................. 213

CHAPTER XII. *Plains and Lowlands.*—Impressions received from lines—Valley silence—Echoes and reverberations—Valley shadows—Sunset valleys—The age of the valley—The brook again—Valleys in autumn and in winter—The valley home—The table-lands—In Montana—The Bad Lands—Colors of decay—Plateaus and steppes—The primeval tracts—The American prairie—Prairie fires—Treeless tracts—The roll of the divides and swales—Prairie wildness—Nature's revenges—The wilderness again—Flat plains—Low-lying tracts by sea or river—The livable lands—Sky and horizon once more—The marshes and meadows—Reeds and rushes—Flags—Beauty of the commonplace—The marsh landscape—Near to civilization—The bottom-lands—Swamps and jungles.................................... 235

CHAPTER XIII. *Leaf and Branch.*—The New World vegetation—The foliage in America—Timber growths—Variety of forests—Depths of the timber—The "Big Woods"—Botanical classes of trees—Tree characteristics—Tree forms—Branch ramifications—The pathetic fallacy—The so-called sentiment of trees—Life of the oak—Tree motion—Sounding-trees—Leaves in motion—Trees in storm—Winds in the forest—Bare boughs—In March—The March harmony—Warming color—The budding season—Summer foliage—Variety of the greens—Light transformations—Swift color-changes—The trees in blossom—Blossom storms—Autumn glory—Indian summer—The scarlet foliage—Harmony of the scarlet landscape—Nature's sacrifices—Tree contrasts—Tropical forests—American forests—European woodlands 253

CHAPTER XIV. *Earth Coverings.*—Trees and shrubs—Bush growths—The substitutes of nature—Laurel and rhododendron—California chapparal—Sage brush—Upland bushes—Common growths—Wild roses—Growths under shadow—Fern and bracken—Scotch heather—Heather color—Golden-rod—Blue asters—Bushes and flags—Meadow growths—The grasses—The earth-protectors—Meadow and pasture—The natural *vs.* the artificial—Meadow flowers—The wealth of color—Pasture changes—Nature's care—Cultivated growths—House and lawn flowers—The mosses—Moss structure—Moss colors and textures—Gray lichens—Rock-staining by lichens—The work of the mosses and lichens—Heat, light, and moisture—Nature immortal—The Great Peace.. 278

NATURE
FOR ITS OWN SAKE

NATURE FOR ITS OWN SAKE

CHAPTER I

PURE AND REFLECTED LIGHT

A FISH at home under some ledge of rock in the depths of the sea, what does it know of sunlight? Doubtless the pupils of its eyes contract and expand with the lights and shadows that break across the hills and valleys of the ocean world, but how dim must be those lights, how densely dark those shadows! A ray of sunshine passing through five hundred feet of water is broken, deflected, almost extinguished; and the eyes that look upward toward the light through that great green lens of wave can gather but a faint glimmer of the truth. They are focused for the ocean depths, and when the fish is brought up to the open day the eyes are instantly set, and stare without meaning. The first flashing sunbeam doubtless shocks them senseless. The truth when revealed is blinding, and our sunlight is final truth to the fish.

Knowledge of light.

We have, perhaps, a contempt for the knowledge of light possessed by the inhabitants of the deep, but our contempt is somewhat shallow. For we ourselves are living at the bottom of an even greater sea—the vast atmospheric ocean. We are looking up to the light through countless strata of air that break and twist and shatter the sunbeam—looking up not through five hundred feet, but probably five hundred miles of air-wave. Perhaps, were we brought up and out of our sea and into the regions of space, our eyes, too, might be blinded by the sharp shaft of a pure and clear sunlight. Our knowledge of it is only comparative, a step upward from that of the fish. The truth in the superlative degree will never be attained. Human eyes have never seen pure sunlight, and that white light which we regard as such is anything but pure. It is not the sum of all radiation, as we are accustomed to think, but the residue, that which remains after the passage through atmosphere.

Atmospheric obstructions.

The air we breathe is filled with countless particles of dust, smoke, soot, salt crystals, vapor; and these particles break light into color by obstructing the beams. The sun ray is thus disintegrated as soon as it encounters our outer atmosphere. Some of it is practically

Particles in the air.

lost to us in the upper air, and that which finally comes on down to the earth has to our eyes a prevailing whitish, reddish, or yellow tone, dependent upon the density of the air. If we could sweep away our atmosphere entirely, the light would appear bluish and the sun itself violet-blue.* There is a predominance of violet and blue in sunlight, but the waves of these colors being the shortest and weakest in travelling power, are the first ones to be caught and absorbed by the upper atmosphere. Held in check, entangled as it were, quantities of them are massed above us, making what we call "the blue sky." The yellow and red waves, having greater length and power than the blue ones, penetrate the atmosphere deeper and come to us with the tale that the sun is yellow or red or, in combination with other colors, white.

Violet-blue sunlight.

But the tale is deceptive. Sunlight in its entirety appears whiter and then bluer, in proportion as we rid ourselves of our atmospheric lens; and the sky itself grows darker from the non-diffusion of the sun's rays. An ordinary rain-storm that clears the atmosphere will tem-

---

* This is the conclusion of Professors Langley, Young, and other scientists. If seen from a distant world, our sun would appear as one of the blue stars.

porarily make the sky and distant hills look bluer, the sun whiter, the light purer. Cold that is intense enough to rid the air of moisture will also make a noticeable difference in the quality of the light. In Manitoba, where the thermometer often sinks many degrees below zero, a bright winter day reveals an air the moisture of which is frozen into floating crystals of hoar-frost, the sky appears cobalt-blue, the sun is white, and when it rises in the morning it is accompanied by two sun-dogs or parhelia, one on each side, and almost as brilliant as the sun itself. The result is a bewildering display of white light that borders upon blue. Every snow crystal glitters, the cup of the sky seems to be lifted into infinite space, the snow shadows are intensely blue, and the running waters are dark-purple in hue.

As we rise above the denser strata of atmosphere that lie along the earth, by ascending mountain heights or otherwise, the light changes even more positively. From the top of Mt. Blanc the stars are seen at midday shining upon a dark blue-violet field that extends down to the horizon; from Pike's Peak the sky is seen to be of a violet hue at times, and not infrequently blue-black; and from Mt. Whitney

Professor Langley observed the sun go down, not gorgeous in color, but coldly luminous, with the dark sky crowded close up to the disk, and the zenith deep violet-blue. Whenever or however the thickness of air between us and the sun is decreased, the coloring of light changes, growing from a yellow flame somewhat like candle-light to something kindred to the blue-violet flame of the electric arc-lamp.

The atmosphere then is chiefly responsible for the quality of our light, and upon the clearness or thickness of the atmosphere depends also the quality of our coloring. If the air is comparatively clear, the light will be sharp and the prevailing notes of color in landscape will be blue and green, because the slightness of the interfering media allows the short color-waves of blue and green to come on down to earth in great quantities; if the air is heavy with particles, the light will be less intense and the notes of landscape will be yellow or red, because the density of the interfering media allows the stronger color-waves of yellow and red to pass through and down to earth, but obstructs the blues and greens. It is owing to density of atmosphere that the heated portions of the globe, like Morocco, for instance, are less strong in

Earth coloring regulated by the density of air.

coloring than the temperate New England, notwithstanding the intensity and the directness of the sun's rays near the equator. The heat of the equatorial region produces dryness of the soil, and dryness produces dust, which is carried up into the air by rising currents. This obscures and changes the color of light more effectually than perhaps we realize. Professor Langley tells us that from the top of Mt. Whitney he saw this dust lying below him like a great reddish mist suspended four or five thousand feet above the level of the surrounding country. It can be imagined that light streaming through such a mist must be not only obscured, but must give a coloring to the earth of yellow, orange, and red, somewhat as the coloring of a room is affected by red or amber glass placed in the windows.

A practical illustration of a dust-laden atmosphere and its color effect was shown us in 1883. The volcanic eruption of Krakatoa threw a shaft of fine ashes some eighteen miles directly into the air, where it was caught by the winds, and swept around the globe; and for months this fine ash was slowly settling through the atmosphere to the earth again. The result was a turbid air and an extraordinary series of red dawns and sunsets seen in many lands. In Spain, where I happened to be a year later, the dawns were the most ruddy I have ever witnessed; and each night the sun went down hissing hot into the Atlantic like a ship on fire, throwing great flaming signals of distress far up the zenith as it sank.

But while the dust veil may produce great mass and variety of colors, these are not necessarily of the highest intensity. The most brilliant hues are to be seen where the light falls the clearest, and this is not in the heated tropics, but near the cold poles. The northern countries have not the many local colors of the tropical lands, but those they possess have more depth and clearness. No palm on the banks of the Nile ever had such brightness of greens as the pine and the spruce on the Norwegian mountains. In upper Scandinavia the flowers are brighter, the sky and water deeper blue, the mountains purer purple, the sunsets more scarlet than in Italy, Greece, or Algiers. And we all know what report the arctic explorers have brought back to us of brilliant skies, flaming Northern Lights, and intense blues in water, ice, and snow seen in the polar regions. There is not the slightest reason to doubt the truth of the

Color in the tropics and at the north.

Arctic colors.

report. Theory and observation both confirm it. A red, a blue, or a green at the north is harsh, intense; where near the equator it is slightly bleached or blended with other colors by reflection. That the latter is more harmonious than the former is quite aside from the present tale.

Sunlight in summer.

The changes in color and light, and their effect upon the world about us, are things of which many of us living in the temperate climes have small appreciation. Our conventional remark to a neighbor in passing, "A fine day!" means merely that we find the weather normal and the sun shining. We have never stopped to study the varieties of illumination and hue that weave and interweave through that day. It is merely a glittering generality to us; yet from dawn to dawn how marvellous is the light, how splendid is the coloring of a clear day in summer! It usually begins with the faint graying of the eastern sky above the horizon, or it may be that the light appears at first high up in the sky. The air has been cooled and somewhat cleared by the night just past, moisture is more predominant than dust, and the consequent sky-color is gray or silver. The light soon extends down and around an eighth of the horizon circle, and then perhaps to a sixth of it; or it may mount upward in the shape of a fan. Sometimes pale yellow is a predominant coloring, and in warm weather a rose hue is quite frequently shown. If the sky above the horizon is barred or streaked with clouds, almost any conceivable color may be reflected from them, dependent upon the state of the atmosphere and the position of the clouds. Again, if the air is dense with vapor or dust, the advance arms of the sun may be seen reaching far over the night like the silver shafts of an enormous searchlight.

The dawn.

These premonitory signs of the coming day are often extraordinary in their appearances. For instance, in Egypt, during the heated season, the dawn is not always the slow stealing of light along the horizon. On the contrary, a single shaft like the pinion of a wing rises upward toward the zenith. In a moment another shaft begins rising by its side, and then another and another, until the whole half-arch of the heavens resembles two spread wings poised perpendicularly. These are, I imagine, the biblical wings of the morning that fly to the uttermost ends of the earth. At other

The dawn in Egypt.

times the Egyptian dawn shows a mild effect of sun-dogs, such as are frequently seen in cold, snowy lands. In the one case, the parhelia are produced by ice crystals in the air, in the other case by dust crystals in the air. They are more brilliant from ice than from dust, and wherewith the one they centre in great spots of light, with the other they shape themselves into side illuminations that resemble wings spread laterally. These, I imagine again, are the wings of light supporting the golden disk of the sun, that may be seen to this day carved on the temple lintels of ancient Egypt.

Wings of light.

But the dawn in our temperate clime is not so unusual in appearance. It is with us the gradual expansion and intensifying of radiance. The light is a soft, lustrous one, illuminating the earth entirely by reflection. While the sun is below the horizon no direct rays can possibly reach us. The shafts are shot up against the blue vault, and from this transparent blue of atmosphere they are reflected back to earth. It is not a bright or sharp reflection. The rays are bent and thrown back only by the infinitesimal particles that float in the upper air. Even when the shafts strike a cloud they simply make it glow like a great

The dawn in temperate climes.

pearl, and the glow is infinitely more delicate for its surrounding of translucent atmosphere. Yet the great vault is illumined, and, as the sun rises higher, far to the north and far to the south, half-way around the circle, a tapestry of silver and gold is weaving on a blue-gray ground, and the dark ultramarine of the west turns a shade paler and seems to lift into space as the light grows stronger. How like the flooding of the tide this light drifts up, and in this great aërial ocean bringing with it warmth and color! Soundless and surgeless, rolling in waves too translucent to be seen, rising higher and higher, yet meeting with no ultimate shore, how gloriously it sweeps up and over the world! How swiftly even the "meagre cloddy earth" borrows a splendor from above and reflects the flush of light and color! The mists stir, the trees tremble gently, the dew slips from leaf to stem, and the whole globe seems to awaken from slumber.

The flooding light.

There is nothing more beautiful in all nature than this flooding of light across the sky, across the earth; yet even as we watch it a great change takes place. The sun peers over the horizon and the first beam of light strikes full upon the mountain's highest minaret of

The sunrise.

rock, splashing it with a pale golden hue. At once the hue begins to creep down from the mountain-top, striking the oaks and cedars one by one with yellow shafts until the whole hill-side is mantled with its color. Swiftly the light spreads to the valley, and in a few moments it falls upon the fields and meadows. Immediately begins the phenomenon of light being broken and obstructed by opaque bodies such as hills and trees, and we have the effect of light-and-shade. Immediately, too, the swift vibration of those points of light productive of color is increased, and we have the brilliant hues that mark the earth under sunshine. Every lake and stream and open sea warms in color and glances the image of the sun, and every hill-side and mountain-crag receives the stain of gold. Not the great objects alone, but the infinitely little, the pale wind-flower, the lowly buttercup, the yellow-centred daisy, the tiny violet, the leaf-whorl of the moss, all put on their brightest garments, each one lifting its head to the sun as the great glory of the universe.

Colors under sunlight.

As the sun rises higher the splendor becomes more widely diffused. The color of the rose leaps to a high pitch, the top of the willow is a mass of silver, the poplar seems to shake light from its leaves as though they were trembling little mirrors. By contrast the shadows across the lawn and along the mountain-side seem darker, though in reality they are lighter; and the light itself may seem fainter because widely diffused, whereas it is stronger and fiercer. By ten o'clock the sun is quite high in the heavens. Heat is radiating from the earth. Strata of warm air are forming along the ground, moving uneasily hither and thither in their search for an exit through the colder air to the upper regions. Dust and moisture, too, are rising; and by noon perhaps there is a haze lying along the hills and meadows, the distant valleys look gray and warm in the sunlight, the mountains beyond them are faintly blue, the sky itself looks yellow or rosy. Color is everywhere, more predominant than in the morning, but less contrasted, because the atmosphere has blended and toned all nature to its own golden hue.

The light at noon.

How different this hot light of noon from the dawn-light! The latter is preferred because it is soft and agreeable to the eyes, but it would be difficult to imagine anything more beautiful or more splendid than bright sun-

Beauty of the noon-light.

light beating at midday upon a field of ripened grain where the fiery red of the poppy gleams in between the yellow stalks; or again this same light falling upon fields of golden-rod or upon great masses of variegated autumn foliage. Blinding, too, as is the noon-light upon desert sands or prairie uplands or flat smooth seas, yet its breadth and intensity make it one of nature's great glories. And how invisibly it cuts through the air! On yonder mountain we should notice falling rain or snow or even a slight thickening of the atmosphere; yet all day long the sunbeams fall upon it and we cannot see them. We see the mark they make on crag and tree, we feel their absence when a cloud shuts out the sun; but that is all.

The fall of light.

As the day wears on, the heat increases. The leaves of the trees and the flowers curl and shrivel, the air rises quivering from the dusty road, the sky grows more rosy—even iridescent. The ascending air-currents are active and the atmospheric particles more numerous. Hour after hour the aërial envelope grows denser and heavier, the shadows fainter, the light more diffused. At last, when the sun has fallen to the western horizon and throws its rays along the surface of the earth, they pass through many miles of this heated dust-and-moisture-laden air. When they reach our eyes they tell the oft-told tale of the brilliant sunset. The pale grays and silvers of the dawn, produced by the sun's rays coming to us through a cleared and cooled atmosphere, have now changed to the golds and scarlets of the evening, produced by the rays coming to us through a heated and a thickened atmosphere. So dense is the air at times that the shafts of the setting sun may be distinctly seen radiating up the sky like the spokes of an enormous fiery wheel; and again at other times the air may be so thick that it obscures the sun's rays, and we can see the red disk go down almost without a flash of light as though its own heat had consumed it. *The sunset.* *Red sun disks.*

The glare and heat of sunset colors are perhaps more apparent than real. The same sun at the time it looks red to us may show the yellow of noon and the white of dawn to the people and the lands lying to the west of us. We are looking from a land of shadow toward one that is still in full sunlight, and the brightness of the sky-color is great by contrast. The colors and combinations of colors that we see on the western sky and clouds at sunset and *Sunset colors.*

twilight hardly admit of description. All hues, all tints are possible, and nothing is of long duration. The appearance is almost as transient as the aurora, for it is shifting in position, shifting in light and color continually. When there are no clouds, the normal evening sky shows a continuous spectrum, and the order of colors begins with red at the horizon and extends in successive bands through orange, yellow, green, and finally shades into the blue of the upper sky. These colors are intensified or depressed by atmospheric conditions, and they are complicated by the appearance of clouds, though the order of their appearance even with clouds is usually maintained, the reds being the lowest down and the succession rising through the intermediate colors to blue.

Spectrum colors on the sky.

The most splendid evening effects are, generally speaking, in the autumn, when with Indian summer there is much heat and dust in the air. Scarlets, carmines, rubies, and burning golds are then apparent. After several days of rain have left a damp, thick atmosphere, a clearing western sky with fleecy clouds will often show very brilliant yellows in bands, and in between these bands small spaces of malachite green. The winter and the early

Sunset sky effects.

spring sometimes show wreaths and scarves of yellow or red upon the clouds after sunset; but as a general rule these are not the seasons for bright displays.

The coming of the dawn and the passing of the sunset doubtless occupy the same length of time, but to us the latter often seems of shorter duration. At the equator there is comparatively no glow on the sky after the sun disappears. Almost immediately upon the vanishing of the disk from view there is darkness. Along the coast of Norway one may see the after-glow upon the sky far into the night; and farther up the coast the sun itself may be seen at midnight. The shape of the globe and the inclination of its axis account for both these appearances. In the temperate zones we have something between the two extremes. The sun for some time after its disappearance from view keeps throwing light from below the horizon upon the upper sky, and thus produces the effect we call twilight. It used to be reckoned that when the sun had fallen eighteen or nineteen degrees below the horizon the twilight ceased entirely; but according to astronomers it ceases whenever a star of the sixth magnitude can be seen in the sky directly overhead. The

Twilight.

first twilight is, however, sometimes followed by a second glow; and after this has passed there is occasionally another light seen in the western sky called the zodiacal light. This usually forms itself in the shape of a pyramid, with its base toward the horizon and its apex extending zenithward along the track the sun has traversed. It is a pale nebulous light, like that of the star clusters called the Milky Way; it appears more frequently in the tropics than in the temperate zones, at dawn as well as at twilight, and is often referred to as the "False Dawn" and the "Wolf's Tail." The cause of its appearance has not yet been satisfactorily explained.

The zodiacal light.

No sooner is the sun gone (at times before it is gone) than the moon comes up beyond the eastern hills, at first rising slowly and then suddenly bursting into view. If the day has been hot and dry the face of the disk is red or deep orange, abnormally large in appearance, and often bulged and misshapen as regards its circle. We are looking at it through that same lower stratum of dense air which has been rising all day from the earth, and is still rising though the sun has set. It is the dense air that gives the abnormal size and the ruddy color. As the orb rises

Moonrise.

higher in the evening sky and gets out of the range of this heavy air lying along the earth, the disk apparently grows smaller and becomes clearer in light. The red and orange fade out, and we see what is called the "yellow moon." It grows still fainter as it rises toward the zenith and the earth's atmosphere clears and cools; and when in the morning hours it sinks into the west, the disk is whitened and apparently shrunk in size. There is little color demonstration as it nears the horizon again. It is cool and silvery, seldom red or yellow, and slips from view usually unnoticed. *The yellow moon.*

Moonlight is, of course, the light of the sun reflected from the moon. It is not reflected from a bright surface like water (there is no water on the moon) but from dull surfaces like rock; and as a result the reflection is many degrees feebler than its cause. Yet the moon has some surface gleam about it and is hardly like an illuminated transparency hung in the air. By comparison with the sun it has no sharp shafts and is so feeble that when sun and moon are both above the horizon the latter attracts no attention whatever; but after the sun has gone down and the moon rising in the east mingles its light with the twilight of the west, it makes a *Twilight and moonlight blended.*

decided impression on the landscape. The two lights together give us the most charming illumination imaginable. The expiring fire of the one and the soft glow of the other mingle in a strange amalgam; and a lustrous light envelops the world as tender and as lovely as that reflected from mother-of-pearl. There is neither deep shadow nor sharp color; and around the great ring of the horizon, stealing far up the sky, there is a vast blend and mystery of color.

Horizon hues at twilight.

The molten golds and garnets of the west as they steal along the horizon circle to the north and south, change into opalescent tints of yellow, rose, and amethyst; and the blue and silver of the east as they spread out to meet the flush of the west, pass through all the shades of gray, mauve, and lilac. For producing delicate tints of color there is no such light as this double illumination coming from the east and the west. Wonderful in their variety, more wonderful in their unity, these tints drape the whole circle of the horizon like a celestial tapestry. Never for a moment are they fixed or permanent. The great waves of light that came up the blue vault at dawn have calmed down to gentle undulations, but they still heave and roll along the horizon-walls, and at every heave some beautiful combination of color breaks and disappears, some equally beautiful one takes its place.

And when the sun and its cloud coloring have gone, when the moon is not in our quarter, then falls the night shadow upon the earth and through it the shining of the stars. They, too, are affected in appearance by the density or the clarity of the air through which they are seen. The night sky hanging over Sahara is usually a very dark purple, but the stars do not shine brightly upon it, and they have no marked colorings; yet they appear very near, as though one might reach them with an arrow. Where the air is more transparent, as in the north of America, the night sky is deeper, the stars sparkle and throw out tiny shafts of light, and they show to the eye different hues of emerald, topaz, amethyst, ruby; but they do not appear to be at all near us. Jewels shining through a dusky veil, they have but little light, and that in such small points that the impression upon the great mass of shadow lying across the earth is not great. We are able to see about us on a starry night, but is it by the light of the stars alone that we see? Is that light sufficient to illumine the world even in a feeble way? At night one-half of the globe is shut

Starlight.

Star colors.

out from the direct light of the sun, and though far above the shadow, above the atmospheric arch we call the sky, the light streams through the realms of space, yet it leaves no visible track, no illumination, no reflection. Beyond our sky it is supposed there is no air, no vapor, no dust to catch and to reflect light. In space the sun's rays travel direct with no diffusion, no halo, no radiation; and could we see the sun itself it would appear as an intensely bright disk without shafts. It would seem then that, with sunlight and moonlight cut off, we gain little or no light from the upper regions of space, save that which comes from the stars. It is possible that our upper atmosphere may be illumined by reflected sun rays or moon rays, and that thus the light of the stars is helped out. And it is possible, too, that there is something of stored-up light or electrical phenomena to add to the night illumination. These accessories may aid the light of the stars somewhat, but they decrease—the total illumination decreases—as the night wears on and out, and the darkest hour is just before dawn.

So much for the direct and reflected lights of a summer's day. It is one day out of three hundred and sixty-five, and has been de-

Darkness of space.

Other lights of the night.

scribed only in its general features. There are no two days in the year just alike, nor will you ever find one day paralleled or repeated in another day. There is a warmth of coloring and light in midsummer and autumn, a bleaching of hues in the spring, a coldness of light in winter; but these again are only general characteristics of the seasons, and do not indicate the infinite changes in each separate day. The variety of combinations made by nature can never be tabulated or classified. Night after night one may watch the moon rise—watch it riding through clouds, first a dull disk, and then a growing light as it nears the edge of a cloud—but the same effect is never repeated; never the same moon, never the same clouds, air, and coloring. The sun comes up, the sun goes down; but each morning light sets a different glory upon the eastern sky, and each evening light reveals new iris hues upon the burning western clouds.

And so with a different radiance for each hour the splendor of the world goes round, night following day, hemispheres of shadow alternating with hemispheres of light. As the earth turns, midnight and noonday slip over its surface. Revolving around the sun in a slightly

The great variety.

The countless changes.

The whirling world. erratic orbit, flinging off heat or cold as the inclination of its axis to the ecliptic, it follows necessarily that the earth must be continually changing in light and color. There shall never be any monotony so long as the sun lasts and the world spins; and that light which was created on the earliest day is to this latest time the most varied and the most wonderful beauty of the universe.

CHAPTER II

BROKEN AND SHADED LIGHT

ALL the lights that come from the sky and reach the earth, whether from sun, moon, or stars, are broken lights in the sense that they are somewhat shattered by passing through atmosphere. None of them reaches us in its purity; yet, comparatively speaking, we say that sunlight is direct light, moonlight is reflected light, and cloud light is broken light. A cloud between the sun and the earth is merely the interposition of a visible atmosphere dense with particles of moisture, but it has a very decided effect in subduing the intensity of light and darkening the earth. The more vapor-laden the cloud and the thicker through its mass, the darker it will appear and the feebler will be the light filtered through it. If it is a large cloud it will appear, perhaps, unusually dark to us, for the reason that we can see only its shadowed base. On its upper part or top it is, of course, shining white in the sunlight, like the cumulus of a summer day; for a cloud

Cloud light.

will have its light-and-shade like any other object, and the dark massed nimbus, which we call the rain cloud, is not very different from other clouds, save that its base is deeper sunk in shadow.

The lowery day.

The gray, lowery day, so often seen in spring and winter, shows us cloud forms so closely packed together that they make a continuous curtain across the sky, through which light passes to the earth in a neutral but widely diffused illumination. This is broken light in its most positive form. Dispersed in every ray by moisture particles, the crippled sunlight can do no more than throw a gray monotone over the face of nature, taking the cloud coloring for its chief note. Such a day is usually declared "dull." The sky and sun are completely shut out, there is no sharp flash of light, color, or shadow, no mellow haze upon the earth, no gilding and fretting of gold overhead. The cloud curtain covers the sky and draws down below the horizon-ring like a cap, a film of mist lies across the meadows, blue and purple drifts of air float high up in the valleys, and along the mountain-sides and over the craggy peaks hang gray fringes of rain. Upon days like these the clouds troop on across the sky, rank upon rank, one so close upon the heels of the other that they are scarcely to be distinguished. How often the traveller has seen them in Paris swaying above the Arc de Triomphe and drifting down over the Champs Elysées, flooding the city with torrents of rain! How often he has seen them defiling over the plains of Bavaria, covering the Bohemian forests, or muffling the hill-tops of New England! There is no break in the lines, no sunlight streaming through. At times a company seems to lift and lighten and the horizon appears to expand; but it is soon followed by a thicker company, the light darkens, the horizon contracts, and the rain waves through the air like the folds of an enormous mantle shaken out by the wind.

Storm light.

And how dark the night following such a day! There is no moon, and only the sharp-pointed stars illumine the watery canopy from above. On such a night the wind seems to rise as the darkness falls, the mountains fade into vague black spots and then blur out, the breakers with phosphor-white crests fall heavy and booming on the sea-shore, and the forest moans and vibrates like a vast Æolian harp. There is little beauty here, save in sound and contemplation. Not even lightning throws a momen-

Night and storm clouds.

tary flash upon the scene. The swirl and the swish of the elements, especially on the sea or on the plains, the sublimity of the tempest, appeal to us perhaps; but our eyes are almost useless. Nothing so darkens the earth as night and rain clouds under a moonless sky.

Mists and fogs.

It is, apparently, a very different light that we see when the clouds are not above us, but around us. A mist or fog is merely a cloud formed close to the ground, and is not different from the cloud that is about one at times on a mountain-top, except that the fog appears to be more luminous and to have more color. Doubtless something of this appearance is due to the thinness of the bank. It generally forms with a clear sky overhead, and is sometimes not higher above the earth than a house-top, though it is often a hundred or more feet in thickness. When the bank is shallow we are surrounded by diffused and refracted light, and an upward glance in the direction of the sun shows us a white light seen as through alabaster. This same light is sometimes seen in the early morning illuminating the whole landscape when the fog has lifted a thousand or more feet above the earth and is spread out into a thin, gauze-like sheet. The thinness of the sheet prevents obscurity and facilitates diffusion, as does a ground-glass globe upon a lamp. The result is a vapor-like light of marvellous luminosity and great beauty. Unfortunately it is not of long duration, and here in America it is not often seen. In France along the Seine, in England along the southern coast, and in Japan it is of common occurrence. The so-called "white horizon" results from a similar set of circumstances. The vapor-laden atmosphere of the morning, seen in mass as we look toward the horizon, produces the white-light effect. Seen in the afternoon or at sunset, the same horizon shows rose, lilac, or mauve tints, because the vapor particles have been superseded, or at least alloyed by the dust particles, and the heat is greater.

But to return to the fog along the ground, as soon as it begins to lift it becomes lighter and brighter until finally the sun peering through from above appears as a silver or pale-yellow disk without radiant shafts. The light grows more golden as the fog-bank decreases in thickness, until at last, the sun having burned its way through to the earth, we see the normal light of day. The fog then disperses in small patches, is evaporated and carried upward by rising currents of air, and in a short time has disappeared entirely. Of course the deepening and the thickening of the fog-bank enfeeble and gray the light. When combined with dust and smoke, as in large cities, it is sometimes dense enough to require the lighting of street lamps in the middle of the day. How it obscures the vision everyone knows who has been in London at such times, or has crossed on the New York ferry-boats, with the pilots picking their way by the sound of whistles and bells. In such fogs a few feet are often sufficient to efface objects entirely.

In the country a fog never appears to be so thick as in the city, though in low marsh places it banks up and obscures land and water very effectually. Seen from a high place looking down, the shore-fog is not unlike a cloud below one in an Alpine valley; and with the sunlight beating upon it the fleecy spun-silver effect is just as beautiful on the one as on the other. There is no limit to the fantastic forms a fog will assume when seen from a height. At times when the dark tree-tops protrude above it the appearance is that of a landscape buried in snow, at other times the meadows seem flooded with milk-white water, or suffocated with drifts and currents of smoke. The small islands off the coast of Maine are remarkable for fog effects, and in cold weather, when the fog turns the bare trees into traceries of frozen silver, the effect is truly splendid.

But close contact with fogs in either city streets or country lanes is not a thing enjoyed by the average person. People grumble and cough and talk about "disagreeable" and "horrible" weather, but not one out of a hundred gets his head far enough out of his coat-collar to see the beautiful pearl-gray tints about him. Broken and obscured as the light is, it still comes through in minute reflecting points. There is nothing opaque about the bank. It is luminous always; and though we think of it and speak of it as gray and monotonous in color, we have only to contrast it with engine steam to find that it is often full of delicate pinks, lilacs, and pale yellows, especially when it is lifting. These minor broken color-notes seldom attract our attention, and yet they are perhaps as refined tones as we shall find in nature's gamut, if we except the notes of the upper sky at dawn. It is curious that people do not see them, and still more curious that they fail to appreciate them when they are pointed out. The average person is quick enough to remark the red flame

Color in fog banks

Delicate hues in nature.

of sunset, but he seldom sees the dove-colors and steel-blues that lie back of him in the east; he sees a scarlet maple or an orange stain upon a hillside meadow in October, but he overlooks the silvery sheen of the wind-swept poplar, or the cloud-like surface of the Indian grass; he is not blind to Niagara and the Alps, and all the "big things," but he has an unhappy way of never regarding anything that is not "big," and hence loses a great deal of pleasure in life which comes from discovering and enjoying the beauty of the so-called commonplace.

Direct light does not necessarily mean a perfectly clear sky, nor broken light a completely clouded one. There are days of alternate sunlight and cloud light; and indeed, a blue sky with drifting clouds is more frequently seen than any other. The heavy cumuli that lie along the horizon like distant mountain-ranges with snowy summits are not very noticeable as makers of shadow, nor are the thin clouds stretched in strata across the upper zenith productive of anything but a general veiling of the light. It is the thick, ragged, or round cloud, drifting across the sky in flocks, that makes the sunlight come and go upon the earth. When each of these moving clouds is surrounded by a

Alternate light and shade.

field of blue the shadow of the cloud is cast upon the earth in isolated silhouette. As the cloud moves, the shadow moves too, and we have that charming effect called the flying shadow. If there is a stiff wind blowing and the clouds are closely packed together with only loopholes of blue here and there, or if the clouds are long rolls of the nimbus with occasionally breaks in the line through which the sunlight falls, we then see that other charming effect called the sun-burst.

The sun-burst is often seen in summer weather, especially if the day is hot, and the air is heavy with dust and moisture. Under such conditions the bright beam thrust through a cloud opening makes a Jacob's ladder of light from heaven to earth. The light falls in a shaft very much as the pinion of the Egyptian dawn rises toward the zenith, except that it is usually frailer and more golden in hue. And it always falls through the shadow cast by clouds just as a beam of sunlight flashes into a darkened room and is seen because it is surrounded by darkness. When a cloud passes across the face of the sun its edges may turn to molten silver and its thicker portions glow with light, yet the beam does not get through and the falling shaft

The sun burst.

is not seen; but just as soon as a flash from the sun breaks through a torn portion of the cloud, the shaft falls to earth and is apparent from its shadowy envelope. It appears to fall earthward in a straight line, but, like all sunbeams, it in reality describes a curve through the lower atmosphere, especially if the sun is low in the heavens. The trajectory is not point-blank, but falls short like a spent rifle ball. Yet this is not seen by the eye and is known only to scientific calculation. To all appearances the shaft falls straight and remains fixed. It is the shadow of the cloud that glides across the meadows, up the valleys, and over the mountains; the sun-shaft does not shift except where it falls more obliquely as the earth rotates from west to east, or its direction is changed by cloud breaks.

The fall of sunbeams.

The sun-burst is perhaps seen more frequently during showery weather or with thunder-storms than at other times, and it is usually more luminous after than before a rainfall. As the first-comers of the storm-clouds begin to cover the sun, the shaft is often seen in a yellow beam falling diagonally toward the earth. When the shower is passing and the sunlight begins to show again, the shaft reappears frequently in

Sun-shafts and rain.

the form of a white beam, stronger than the yellow one, because falling through denser moisture. There may be many of these shafts, and they may radiate in all directions from the sun, as one often sees at evening, when the west is barred or streaked with clouds. The reaching down of sun-shafts toward the earth, with or without a shower, is commonly referred to as the sun "drawing water." It is really the sun illuminating the dust or moisture in the air, just as the rainbow, which spans the opposite heavens from the sun, is but the sun's rays reflected and refracted in prismatic colors from drops of rain.

The sun "drawing water."

For variety in the display of sun-bursts I know of no country more interesting than Scotland. In stormy weather at sunset the light falling through chinks of the clouds will often make a half-wheel or fan-shaped alternation of light and shadow most brilliant in its flashes of gray and silver. And again, I have never seen such effects of sun-bursts and flying shadows together as in the Grampians, particularly those more barren portions of the hills where the heather is absent and only a yellow-green of grass and a slate-gray of stone are seen a background. Over the slopes and down the

Sun-bursts in Scotland

valleys the lights and shadows seem to wave in bands, like the streamers of the Northern Lights across the sky. The shaking shimmering effect and the alternate colorings of yellow, green, and gray, chasing each other across hill and dale, are most extraordinary in appearance. After watching them for a few moments, it is quite impossible for the eye to tell whether the light, the shadow, or the color is flying. At other times, when the clouds are rounder and larger, their shadows slip along majestically from crag to lake, from lake to crag again, gliding noiselessly and without obstruction up and down and over the Scottish moors like dark peering spirits seeking a hiding-place and never finding it. They roam restlessly on and on, until at last they spread out upon the flat North Sea and their dark forms, changed to lilac in hue, go slipping over the waters to the east, still restless, still noiseless, still flying. In other lands the shadow is interesting to watch as it glides across the meadows covered with buttercups and daisies, and climbs the wooded mountains to vanish over the ridge; but the bare hills and moors of Scotland always seem the best playgrounds for the sun-burst and the flying shadow.

Light beams and flying shadows are some-

Flying shadows on the moors.

times seen under moonlight, but they are not so marked as those produced by the sun, because of their want of definition. The moonburst attracts little attention on the land; and on the sea, where there is reflection from a ruffled surface, the spot made by falling light is apparent enough, but seldom the shaft itself. The light is oftenest seen far out upon the horizon, and is merely a flicker and a sparkle upon the water. As for the flying shadows of clouds at night, they are dark purple in tone and are sometimes weird in shape, but unless the night is very bright, they are not usually noticed.

Moon-bursts and moon shadows.

Shaded light is somewhat different from broken or clouded light. It is not produced by shattered parts of direct rays that steal through vapors and cloud-veils, but by widely diffused or reflected rays. The direct beams are usually cut off by an opaque substance, and the light in the shadow is received from the reflecting sky, the air, or some other illuminating or light-diffusing body at the sides. The earth as a globe is a good illustration of this. It is light on one side, and its opposite side would be absolutely black were it not for such reflecting bodies as the moon, the planets, and possibly

Shaded light.

the illuminated upper atmosphere—counting out for the present the faint if direct light of the stars. Were it possible for a tree or a house to be in the far upper space where there is no air, its sunlit side would be intensely brilliant and its shaded side coal-black; but on the earth the shadow of a tree or house is illuminated by the atmosphere surrounding it, and by the side reflections thrown upon it. It is the diffused light, produced by atmosphere or otherwise, that makes a shadow luminous, and it is the sharp, direct light that makes a shadow dark. One may state a general rule in these terms: The greater the diffusion of light, the greater the expansion and illumination of shadows; the sharper and more direct the light, the more contracted and the darker the shadows.

We can see this well exemplified almost any night by studying the light of the electric arc-lamp. It is the strongest and the most direct artificial light we possess; moreover, it is a white light, with much of blue and violet in it, and the shadows produced by it are very dark and clear-cut. Seen at night, these shadows cast by the bare limbs of a tree upon pavement or upon snow are precisely edged, have little penumbra, and are almost inky in

The law of shadows.

Electric-light shadows.

their blackness. Gas-light will cast no such shadows, nor will the sun, nor will the arc-light itself when muffled by a white globe. Anything like thick atmosphere, a cloud, or a milk-white glass that will spread the light over great space will lighten and expand the shadows at once. Hence it is that on cold, clear days, when there is little dust or vapor in the air to diffuse light, the shadows are darker, sharper, and less noticeable in their coloring than at any other time, while the hot days, with their thick atmospheres, produce opposite results. *Diffusion of light.*

In America the heated days of early autumn, so remarkable for their hazy envelope of air and bright coloring, produce odd changes in the illumination of almost everything in landscape. The shadows become much frailer in body, more transparent in light, with very pronounced hues, especially in the tones of lilac and blue. During the three heated days of September, in 1895, I had the opportunity of studying color effects, in both light and shade, in the woods and fields near Princeton, New Jersey—one of the most brilliant spots in autumn I have ever known. The studies were interesting, but the material was so bewildering in variety that I found great difficulty in *Shadows in hot weather* locating causes and arriving at conclusions. The trees, the bushes, the field grasses were already tinged with autumn hues, and these hues, enhanced by the heat, made the landscape appear crude and violent in its coloring. No imaginable tint was absent from the scene, and the greens, reds, yellows, and oranges were flaring in their intensities. But what impressed me more than anything else was the iridescent coloring of the atmosphere, the wavering of the heated air, the faintness of the shadows and their pronounced body of color. The prevailing tints in the shadows were lilac, violet, and rose. There were few shadows that were colorless, and few, if any, wherein the local color of the ground or object they fell upon was not twisted or distorted somewhat by a reflected or a complementary color.

Colored shadows.

It is not a new theory of science that every color casts its complementary hue in shadow. The practical working of it may be frequently observed in nature. A sheet of white paper catching the light from a red sunset will receive a green shadow from an object interposed between the paper and the sun. The same red light of sunset falling upon snow will sometimes produce green in the shadows of trees

Scientific theory.

and bushes. Dr. Weir Mitchell has noted yachts at sea sailing in the track of a fiery red sun with the shadowed white sails showing "a vivid green;"* and I have seen more than once the white sails of yachts crossing a yellow sunset when the change to blue in the sails was strongly marked—blue being the complementary color of yellow as green is of red. Undoubtedly the yellow sky at sunset is measurably responsible for the blues and purples of the mountains below it, and the more intense the yellow the stronger the blue-purple. If the sunset shows greenish-yellow, the mountain shadows will be violet; if orange, the shadows will be cyan-blue; and so on throughout the gamut each color will disclose its opposite in shadow.

Complementary hues in shadow.

This is scientific theory, and it has been demonstrated and proved true of nature when all the conditions are just right. The only trouble is the conditions in nature are seldom just right. The complementary coloring in the shadow is apparent only on certain days, and under certain lights, atmospheres, and temperatures. It is an error to suppose that a color is always casting its complementary hue

* *Doctor and Patient*, page 176.

in shadow or, at least, an error to suppose that it is always apparent to us. There are influences, too, such as the local color of the ground and the sky reflection, that may neutralize or utterly destroy the complementary hue. It might be thought that a yellow sun at midday would produce blue shadows under the green maple on the lawn, but as a matter of fact it does not. The color of the shadow, whatever it may be, is absorbed and lost in the green of the lawn upon which it falls. The same tree shadow falling on pale-gray clay, or across a sandy road, will show blue or lilac at once; but I do not think this is owing necessarily to the presence of the complementary hue. It is more likely caused by sky reflection, helped out, perhaps, by atmospheric reflections from the sides.

The blue shadows upon snow, so common in winter, are never seen except under a blue sky; and the bluer the sky the more apparent the blue in the shadow. They are produced by sky reflection, and the sky coloring is faintly apparent on the snow in full sunlight, but more obvious, of course, in the shadow. These blue shadows are stronger at sunrise and at sunset than at any other time. Under a clouded sky they disappear entirely, and only a gray effect is apparent. Just before dusk, when sometimes the clouds become empurpled, the whole body of snow will take on a purple reflection. The same or a similar effect is noticeable in the sand dunes along the sea-shore, though sand is perhaps not so good a reflector as snow. I should account for the lilac shadow on the clay or broken-stone road in the same way. It is a mingling of local color with sky reflection and side lights rather than complementary hue. A rough surface like a green lawn or a meadow will not show a colored shadow at any time or under any conditions, so far as my observation goes; and I believe the reason for it is that it has not a favorable surface for reflection. *Lilac shadows on clay and sand.*

If colors were always pure, and if side lights, atmospheres, and sky reflections could be eliminated, we should undoubtedly see the scientific theory of complementary colors always demonstrated in nature; but the problem is complicated, and all talk about "pure colors" is misleading. Nothing is pure; everything is mixed and alloyed. The neutralizing effect of side lights, complementary and reflected hues, and local grounds, puts scientific calculation out *Mixed colors in nature.*

of countenance. A pure color in nature is always more or less bleached, grayed, silvered, or gilded—changed at least from its original estate — by these conditions. What might be the green of a maple-tree lighted by sunlight alone is one thing ; what it is lighted by sunlight, sky-light, and reflected light from the earth, not to mention atmospheric influence, is quite another thing. When all the factors are considered, we have anything but a pure green in the tree. It is, doubtless, a mingling of many hues that favors the mauve, the rose, and the lilac shadows. But then, again, they seldom appear unless the day is hot and the air thick, which leads one to think that atmospheric reflection plays some part in their production. The cause can be conjectured only, but there is no doubt about the effect. The colored shadow is a reality, though its recent discovery finds people still somewhat sceptical about it.

Shadow complications.

We have seen that clear light is favorable to the sharp-cut shadow, and that when the light is more widely diffused by atmosphere, or increased by reflection, the shadow begins to lighten, to become vague and soft on the edges, and to be enveloped by a penumbra. When the light is still more widely diffused and broken by coming through clouds, it is commonly supposed that the shadow disappears entirely. We think of a cloudy day as a shadowless day, and practically it is so. The outlines of the shadow are lost, and yet the shadow itself is there, if we will but look for it. The green maple on the lawn has its breaks of light and dark seen in the foliage, and its form is cast in shadow on the ground, but the latter is very faint. It is only by the generally darkened tone that we can detect the shadow on such a day, and even then there is little distinction in color between it and its surroundings. Sometimes at a distance the shadow will appear bluish, but that effect is atmospheric rather than reflective. Sometimes, too, odd colors will creep into the shadows when the sky overhead is clouded and there are spots or breaks of light along the horizon; but when the whole sky is under a veil of cloud, the color of the shadow is practically neutralized, and takes its hue from the ground upon which it is cast.

The shadowless day.

Odd colors in shadow.

The conditions of shadow production under moonlight are similar to those under sunlight, except that the degree of both light and shade is largely reduced. That the direct moonlight produces color wherever it strikes the garment-

Moon shadows.

ing of nature is undoubtedly true, but it is always a subdued dull color. And the shadows, though they are luminous and not black opaque patches, have only dull shades of blue, purple, and gray. There is a modern tendency to see too much color in moonlight—in fact, to see more than really exists. The old idea of the whiteness of its light and the blackness of its shadows has passed away, but the new idea has some extravagance about it. Colors of every kind under the moon are far removed from the feeblest of daylight tintings.

Star shadows.

Feebler still than the moonlight is the light that comes from the stars. The planet Venus and many of the fixed stars are bright enough to throw at times a long reflecting track upon ruffled water, but the colors produced by them upon landscape are blurred into smudges of dark purple and blue, and the hues of the shadows are too vague to be seen.

CHAPTER III

THE BLUE SKY

THE two great expanses, the blue ocean at our feet and the blue sky over our heads, are both impressive in vastness—the ocean more than the sky, possibly because we are familiar with its extent and have felt its power. We know, in a vague way, that the sky is even vaster than the sea, that it is the open field leading into interminable space; but its very obvious coloring, its apparent arch on all sides springing upward and inward from the horizon, its fixity, give us something of a false impression. We are inclined to regard it as a great blue dome or roof, a something tangible that is supported by the horizon-rim, a concave surface looked *at* instead of a vast transparency looked *through*.

Impressions of the sky.

And there is some excuse for our regarding the blue sky as an actual surface. It is the outer envelope of the globe, and is made up of the blue rays of the sun reflected from atmospheric

Transparency of the blue.

particles. These reflecting particles seen in mass apparently make a roof above us which looks to be ten or fifteen miles in height. It is merely an appearance, however, and our not too reliable eyes deceive us. It is known that the atmosphere is from two hundred to five hundred miles in thickness, perhaps more, and there is no demarcation line where the blue begins or ends. Nor is there any point in this blue where cloudiness, haziness, or opacity shows. There is not a blur or film upon it, save where it is influenced by earthly vapors and dust. The sky itself is everywhere transparent, else we should not receive light through it or see the sun, moon, and stars shining beyond it.

Sky depth.

The recognition of sky distances is not easily made by the eye. A glance upward may tell us of five or fifty miles, as our imagination rather than our focus is adjusted. Looking out and over a tract of earth, we conceive distance by perspective lines, by objects decreasing in size, by the diminution of color, and the increased thickness of atmosphere. They are all optical guide-posts by which we can reckon with depth and width. But no such conditions exist in looking skyward. It is true we are looking through thick air to thin air, and beyond that into black space, but the color gradations are so subtle that we do not perceive the changes from one to another. Clouds help us somewhat in increasing the feeling of depth, for they are perspective points five or six miles on the way at least. Sometimes a pillar of cumulus will rise in the air thirty thousand feet from base to top, and tracing this upward the eye may see far above it the drift clouds of the stratus, and still higher, like specks upon the blue, the fine-spun fibres of the cirrus. This will give some idea of distance, though it is not entirely satisfactory. The view from Alpine peaks, where we are already twelve thousand feet up, and see still far above us against a violet sky the white spirals of the ice clouds, is not more satisfactory, save that in the thinner and clearer air the feeling of space is greater, and the sky becomes more of a blue wilderness than a domed roof. *Through the clouds.*

We comprehend the breadth and reach of the sky perhaps as little as its depth. Our horizon is an apparent circle as our zenith is an imaginary point. The circle is twenty, fifty, or from high ground perhaps seventy miles in diameter, but we always see its outside limit—the complete circle—no matter how vast the view. Nowhere is the eye so hemmed in, nowhere does *Sky reach.*

Sky lines seen at sea.

the horizon-ring appear so small, as upon the open sea. The ship upon which we stand is the centre of a watery field, the mainmast points overhead to the centre of the blue firmament, and all around spreads the deep azure glow. Judging from vision alone the world appears very small. The uttermost rim is just beyond us. The expanse of the sea and the reach of the atmosphere about the whole globe are practically unfelt. Even the height overhead seems greater than the sweep before and after us. The limitation becomes still more limited when the vapors lying along the surface of the sea thicken the air and obscure the sight. We cannot as a general rule under favorable conditions see more than fifteen or twenty miles across sea water, and even in calm weather the horizon is often clouded by vapor banks that lie along it like a row of faintly seen hills. All this helps the illusion of being circled and shut in by the horizon. Then again the sense of distance by perspective lines is practically annihilated. Occasionally the skeleton masts and black trailing smoke of an ocean steamer, or the tower-like looking sails of a square rigged ship appear, and act as catch-points; but these are slight, and as for aërial distance we recognize it only by obscurity of coloring, which at sea dulls the vision instead of clearing it.

It is on the land, and from the mountain-top, that we gain the best idea of the round reach of the sky. From such an elevation we not only see hills and valleys stretching away and down the sweeping world-circle, but if the sky be spattered with the white cirro-cumulus clouds, driving along in flocks before the wind, these, too, will seem to slope outward and downward like the earth. The result is that the impression of expanse in sky and earth is prodigiously enhanced. The view is awe-inspiring; and it is not necessarily so because it belittles the objects directly below us, but because it gives us a larger idea of distance, space, and sweep. The world seems a greater globe, the sky becomes enormous, and the imagination rises to meet the new presentation.

Sky lines seen from heights.

There is no feature of the earth that can be regarded as more fixed, more permanent, than the blue sky overhead. And yet it seems as though a strong wind might blow it away. Winds, however, have small effect upon it. Clouds and storms pass across it, altering and obscuring it to our eyes, but beyond the local disturbance we know the sky is as serene and

Apparent changes in the sky.

unchanged as ever. It never seems to move, it never seems to shift; and yet again, it is far from being an unvarying appearance. Sir Isaac Newton discovered years ago, from the twinkling of the stars and the shaking of shadows cast by high towers, that "the air is in a perpetual tremor." Down close to the ground on a hot day we can see, in little, this tremor of the air as the heat currents rise from the earth; and the mixture and intermixture of hot and cold currents in the upper air, the blowing of winds, and the drift of clouds must shake and disturb the lower layers of the blue, though this disturbance is not often noticed by us. At times I have seen, or fancied I have seen, in studying the clear sky, what might be called waves moving across it. The motion did not seem to be that of ringed waves, such as one sees when a stone is thrown into a pond, but of deep undulations of varying blue succeeding each other slowly like the heave and roll of a glassy sea. Only on very hot days has this effect been apparent; and I would not be certain that it is an actual fact, for the eye after long gazing at light and color is liable to become confused and see falsely. Still, I have seen the appearance a number of times, and I believe it to be reality

Sky waves.

rather than illusion. What causes it I cannot say, but it would seem to belong to some shaking of the lower atmosphere, for I have never seen it from high mountains.

The lower atmosphere is, indeed, responsible for most of the volatile capricious appearances of the sky. From mountain-tops the sky is not so changeable, the stars twinkle less, showing that the atmosphere is quieter, and the face of the blue more uniform and serene. It lies there calm as at creation's dawn, lighted as was the old Mosaic firmament, and studded with the same jewel-like stars. It seems above and beyond all local and temporary disturbances. Winds mark it not, storms are far beneath it, heat, dust, and moisture effect it but slightly. It pales and lightens under the sun, deepens under the moon, and darkens under the stars, but in other respects it shifts not. An enormous sweep of violet-blue, it rests, a type and a symbol of unchanging serenity. *The blue from mountain-tops.*

And oh, the mighty silence of the upper sky! What a contrast it is to the noisy, wind-swept earth and the restless ocean! Infinite realms of violet-blue sweeping outward and upward, yet from them comes only the Great Silence— the hush that tells of limitless space. No *The Great Silence.*

shock, no jar, no clash; there are no hidden spots of earth so silent as the depths where the stars lie buried.

The blue from the valleys.

This perpetual violet-blue glow, unmarred and unspotted by high light or shadow or varying tint, save such as it receives from the sun, might be thought monotonous, did we always have it before us. But humanity does not make its abiding-place on mountain-tops. It prefers the valleys, and there the vapors and earth mists and dust particles produce a different-looking sky from that which is seen from the height of Mt. Blanc. It is fortunate that it is so; yet, even in the valleys, people sometimes complain (it is said that they do in Southern California) of "the monotony of blue sky." In reality the "monotony" is not in the sky, but in the eyes that look at it. Seen through the lower strata of atmosphere, it is never the same for any length of time. Its form is continually changed by clouds and cloud-flocks, new colors are being woven backward, forward, and across it, by shifting masses of atmosphere, its light is waxing and waning with the motion of

By day and by night.

the earth. There is a continuous weave and ravel of delicate-hued textures, and from dawn to dusk there is not a moment's pause. Sun flame shot through, earth reflection shot back, cloud light scattered between, all make their momentary impression; and even at night, though the splendor is diminished, it is not extinguished. The moon lends a pallor to the blue, the Milky Way stretches its nebulous scarf across it, the Belt of Orion blazes out from it, the planets gleam on its dark ground, and through the long dusk of night the shifting splendor falls, the eternal round of beauty moves on. *Shifting splendor.*

And by day or by night, seen from mountains or from valleys, what infinite tenderness in the blue! Was ever depth and transparency of color so beautifully revealed, and by such subtle, elusive means? Drifts upon drifts of air superimposed one upon another, rings upon rings of illuminated atmosphere, rising higher and higher, and all of them deepening the tone, but never clouding its transparency. How far we seem to see into that blue, but there is no place where the eye reaches a background—no place where a basic color appears. It is always a spectral abyss—a blue dream resting above us, which the mind of the human has never been able to grasp as a reality. *Tenderness of the blue.*

It is not to be wondered at that the tenderness of color and the varied hues in the sky are unseen by the average person. I have never met anyone, other than a scientist or a landscape-painter, who could conscientiously say that he had spent five consecutive minutes of his life looking at the blue above him. Its colors are not violent enough, nor its changes swift enough to attract attention. A scarlet cloud draws the eye at once, but the clear sky, with the sun burning a great hole in the blue, and throwing off a ring of pale yellow light that radiates outward, decreasing in the most delicate gradations until lost in the prevailing azure, is scarcely ever remarked. From dawn to dusk pale tints of silver, lilac, and ashes of roses lie all around the horizon-circle, reaching up toward the zenith as though aspiring to be rid of earthly taint; hour after hour the sky overhead is passing from dark blue to pale yellow, from pale yellow to amethyst, from amethyst to opal; evening after evening the cloudless sun goes down, leaving pale bands of spectrum colors on the twilight sky, but all this is waste splendor so far as the average person is concerned. People have an unhappy fashion of seeing with their ears. Someone tells them of the Alpine glow upon the snow-cap of the

Jungfrau and they go there to watch, perhaps days at a time, for its appearance, when they might see the same pink glow upon their own skies at home almost any summer evening. It is not necessary for one to go beyond the dooryard to see beauty. The open sky will reveal more varied lights and colors than anyone could schedule or tabulate or talk about in a lifetime. Seen from our valleys, instead of being a monotonous blue roof above us, it is, perhaps, the most changeable transparency that human eyes have ever looked at or looked through.

Alpine glows at home.

But while this variety is true of any one patch of sky, it does not follow that all blue skies are alike, even in their variety. Atmosphere, upon which so much responsibility for light and color has been thrown, is the potent cause of many different skies over many different lands. In dry countries, where there is much dust in the air, the blue is often a pale turquoise, or if there is great heat, then it is pinkish, or rose-hued. One hears much in tourists' descriptions of "the deep blue sky of Italy," but if they mean by that a *pure* blue sky, their descriptions are not accurate. It is oftener pale lilac, rose-hued, or saffron-tinted, and not to be compared in intensity and purity of blue to the skies

Skies in different lands.

of Scotland. In no warm country is there such clear blue sky as one may see in the northwest of America; and if we may believe the descriptions of Dr. Nansen, the Arctic explorer, this blue grows more intense as we move toward the poles, until at last it becomes of that violet hue seen from mountain-peaks. The Egyptian blue is often "deep" when the air is clear and still, but with winds, heat and dryness it takes on a warm tone as though it were seen through a red dust-veil. A similar effect may be noticed over cities like London, where smoke and soot are continually fouling the air. The blue has a suffusion of pink or copper-color that gives it a hot look. In moist climates like Germany or Holland, there are often very clear skies, but the moisture particles in the air usually tend toward the production of a pale, milky whiteness in the blue. Again, in all countries of the northern temperate zone the purest summer skies are in the months of May and June. After these months the hot and dry summer begins to pale the blue, and in the autumn, when the leaves are changing to gold and scarlet, the sky in perfect harmony becomes rosy and often opalescent.

If people are little observant of the blue sky

Color changes through atmosphere.

Season changes in the blue.

in its color transitions, they are, perhaps, even less observant of its luminosity or light-diffusing power. It is a popular belief that the sky is a screen or veil to the earth, and that its principal reason for existence is that it tempers light to human eyes by obscuring it. And that is partly true. But the blue also receives, diffuses, and transmits light. It is luminous, at times scintillant, in small bright points. By long and attentive watching one can actually see these little points of light twisting, curling, falling and disappearing quickly as though they were mere flashings of star dust. And this does not refer to that portion of the blue sky near the sun where shafts of light are thrown down, but to the portions far removed, which are seen, perhaps, when the sun itself is under a cloud. The pure blue throws out more light than we imagine. If a sheet of white paper be held under it, even when the sun is below the horizon and eliminated from the problem, it will appear much lighter than the sky. But is it lighter? Paper is not a body luminous in itself. All the light there is in it is merely the reflection of what comes from the sky, and a reflection can never be so strong as its original. There is an apparent contradiction just here,

Luminosity of the blue.

Blue sky and white paper.

Transmitted light.

which may, perhaps, be cleared up by some such explanation as this: Glancing up at the sky our eyes look inevitably into the shadows of air particles; the light that comes to us is transmitted through and between the particles. Glancing down at the paper, we are looking into the high lights of the paper instead of shadows; the light is now reflected instead of transmitted.

It is because of the coloring of the blue, and the transmission of light in countless infinitesimal points through it that we fail to appreciate its luminosity, and yet next to the sun and its reflections it is the most luminous phenomenon in the universe. It blinds the light of the stars so that we fail to see them in the daytime, and even the moon looks pale and wan beyond it until the sun has gone down and the light fades out of the atmospheric canopy. Upon

Sky lights on the earth.

the earth its effect is equally apparent. The snow reflects the light of the blue sky like the sheet of paper; and the white daisies of the meadow, the white foam of the sea, and the silver flash from still waters are but reflections of it. From mountain-heights at twilight one may see below in the valley the thread-like river, the white farm-houses, and the fields of yellow grain showing like spots of light upon the shadowed landscape. Whence comes the light thrown back to heaven by these objects if not from the blue sky overhead? Because sky-beams do not fall like rain-drops we think, perhaps, they do not fall at all; but their presence in reflection is about us on every hand.

But possibly more beautiful than the transmission of light is its reflection as shown upon this same blue dome of air. When the sun is in the zenith all the light is transmitted, but when the sun is below the horizon its light is thrown up and under the blue and is reflected. Instead of looking into the shadows of air particles we are looking into their high lights. This gives the effect upon the eastern sky that we call the dawn, and the more gorgeous effect in the west, called twilight. These two effects are the only ones that reveal fully the reflecting power of the sky. If we could rise above the earth and from the moon look out toward this world of ours, we should doubtless see it muffled by a great luminous covering. The light from it would all be reflected and the white, misty air might completely hide the earth from view. It would not, however, be a brilliant or scintillant light. Like that of the dawn, it would be softly pervasive. The atmosphere from which

Reflection from the blue.

the dawn is reflected is not hard or smooth like metal; it is not so compact even as the softest, thinnest cloud of the stratus, yet what beautiful light it throws off! The white light that hangs over a city at night when there is fog, caused by the glare of many lamps thrown back from the fog bank, is brutal and coarse by comparison; and the ruddy sunset caused by dust and cloud is more palpable and less crystalline. There is no glare or flare about the dawn. The light comes from a deep transparency quivering under the rays of the sun, receiving its illumination in straight shafts of fire, and yet reflecting it with a softness of glow that delights the eye by its purity and delicacy.

Atmospheric reflection.

Surely this light of dawn is the highest manifestation of beauty in the universe. Colors do not equal it, lines and forms of cloud and earth are petty compared to it, shadow is its very antithesis. It is not wonderful that it should have been the inspiration of Orphic song and the symbol of deity in ancient religions. To-day it seems a sign of preternatural glory even to modern materialism. Not the sun itself, but its light (symbolic of the purity and luminosity of Deity) bowed the head of Zoroaster.

The dawn light.

The faith is strange with us now, and yet how well founded it was in natural religion. Instinctively all races of men, whether savage or civilized, lift the hands and raise the eyes toward the heavens as though beyond the blue dome rested the seat of final justice, and its shining light was a manifestation of Supreme Power. The spiritual in man has always looked upward and counted the future abiding-place as somewhere beyond that sky; but the light wherewith God "covereth himself as with a garment" is no longer regarded as a token and a message—a call to thanksgiving and to prayer. The muezzin's voice, the angelus bell—some human ritual—now bends the knee where once the white dawn drew all eyes as to the open gate of paradise. In the long centuries of history how many prophets and peoples have gone their way to the grave following symbols of their own making—devices that have turned to dust and mingled with human clay! How many times has the old order changed! How many times have new faiths, new symbols, new signs arisen! Yet the light in the east has never changed, never lost its lustre. Its glory was from the beginning as it shall be to the ending. Modern science may write it down as

In ancient religion.

The dawn as a symbol.

Significance of beauty.

a material phenomenon, and modern creeds may discard its worship as idolatrous; but priest and scientist, in common with all humanity, have felt its splendor and known its beauty. Was beauty then made for ashes, and has splendor no significance? The aspiring soul will not so account them. It believes that He who stretched out the heavens as a curtain and laid the beams of His chambers in the waters makes Himself manifest in the splendor of His light, and in the beauty of its reflection upon the morning sky.

CHAPTER IV

CLOUDS AND CLOUD FORMS

A CLOUD is always a cloud, no matter by what name it may be called or what its form or height above the earth. The fog that knocks about our ears is made up of the same visible vapors as the heaped-up cumulus rising tower-like thousands of feet above us. That one lies along the ground and that the other rises to a lofty altitude is due merely to a difference in temperature and density.

Clouds are formed by sudden lowerings of the temperature of moist air; and this lowering of temperature is usually caused by warm air rising into higher altitudes, expanding as it rises and cooling as it meets with the upper cold-air currents. The simplest and most frequent manner of cloud-making is this: The radiation of heat from the earth forms into a column-like current of air, and the natural tendency of this current is to push upward, seeking an exit into cooler regions. It keeps rising, expanding as it reaches thinner air,

Cloud-making

cooling and becoming moister as it meets with cold currents, until at last it attains a height where the dew-point* is reached. Then begins the change into cloud.

Cloud forms. The hot air of summer rising upward reaches its dew-point very soon, and the usual result is the formation of the large cloud which we call the cumulus. When there is little heat or moisture in the rising air, and no pronounced cold in the aërial regions through which it passes, as is often the case in the spring of the year, the air-current may rise to a greater height, and when finally the dew-point is reached the condensation appears in the form of the stratus or cumulo-stratus cloud. The dryer and colder the ascending current, the higher it must rise before it condenses; and so at times it rises to the region of frost, then freezes into the thin clouds of the upper cirrus, which are made up of tiny ice-needles floating in curls and wisps against the blue sky.

Why clouds float. When once formed, the clouds are heavier than the air in which they float, and their natural tendency from the moment of their formation is downward and earthward. Knowing this fact, we are often led to wonder why they

---

* See Chapter V. for explanation of the dew-point.

do not fall, why they do not rest upon the earth instead of in the air. There are several reasons for their not doing so, and all of these reasons taken together may account for the apparent defiance of the law of gravity.

Thistle-down will speedily find an abiding-place on the ground if there be no wind, but a gentle breeze will carry it drifting for miles, now high, now low, always soaring, sinking, floating. Something of this effect is produced upon the clouds by the winds and the moving currents of air. They are always forming and changing and being kept in motion by the winds. The travelling capacity of the different cloud flocks is, as we shall see hereafter, much greater than is generally supposed.

Effect of the winds.

Another and perhaps more potent cause of certain clouds being kept above us lies in the warm currents of air that are continually rising from the earth and buoying them up, very much as the heated air from a stove or lamp-chimney may buoy up a feather. We can see this illustrated in the formation of the clouds that sometimes hang about a mountain's top. The warm currents of air in the valley seek to rise up the side of the mountain because it is a natural conductor protecting them in measure

Effect of the air-currents.

from sudden gusts of wind and cold. They rush up the mountain-side quite rapidly, as everyone knows who has stood there at noontime and felt the draft upward from the valley. As soon as they reach the top of the mountain they are forced from shelter by the currents coming after, and meet with the cold winds above the peak. The result is quick condensation and the formation of that cloud which is called the "cloud cap" or "night cap" of the mountain. It is broken and blown away by the winds continually, but it is also being continually renewed by the ascending currents, so that apparently it remains stationary and intact. It does not sink down, because of its renewal and because the currents in measure lift it up.

Cloud caps.

Something of the same process is apparent in the formation of what is called the "banner cloud," which appears to fly out like a streamer from some of the Alpine peaks. This cloud is usually on the warm valley-side of the peak. The moist air from below rises along this sheltered side to the tip of the peak before it is struck by the cold currents and condensed into visible vapors. Above it and at the sides the cloud is being cut off and drifted away by

Banner clouds.

the winds. It is visible only where it clings to the lee-side of the peak, and it stretches out into the air as far as shelter is afforded it in the shape of a long, thin flag. At a distance it looks as though it were something permanent, whereas it is only a continuous-forming cloud cut sharp on its sides by the keen edges of the wind.

But these illustrations are of exceptional clouds, and even with them the rising currents alone are hardly sufficient to account for their being sustained in air. The majority of clouds are formed in open space and their air-currents have no mountain-sides to protect them. Nor are the common clouds subject to such violent destruction as the banner clouds. Moist currents are rising, clouds are forming and reforming, changing, sinking, disappearing; but they are not often slashed into strips by the winds. We must seek a third cause for their being sustained in air, and it has been suggested already by the word "renewal." Clouds after they are formed are practically self-renewing. When the ascending air-current condenses into cloud the heat of the air-current goes upward with a tendency to form newer and higher clouds as it rises; but the moisture of the current, robbed

Clouds formed in open space.

Self-renewal of clouds.

of its heat, forms into tiny, cold-water globules which have a tendency to sink down toward the earth. If the globules are large and heavy enough, owing to sudden condensation, they do fall to the earth in the shape of rain; if they are small, as is usually the case, they no sooner sink down into the warmer air from whence they came, than they are evaporated and carried up to the top of the cloud, to be once more condensed into mist. The "renewal" of the cloud then means that the water-globules are continually falling down only to be evaporated and sent up again for recondensation. The cloud is always losing at the bottom, and its flat base shows the line where evaporation takes place; but it is continually adding to itself on the top. The tendency of the cloud at the top is to form above itself drifts of higher clouds, but this is held in check by the loss of moisture, the dryness of the upper air, and the dissipating action of the sun's rays from above; the tendency of the cloud at the bottom is to sink down, but this is held in check by the continual evaporation as the water-globules fall into the warmer, lower air. The cloud then, though in reality always changing, is apparently stationary and without change. The ascend-

Cloud changes.

Recondensation.

ing air-currents feed it, and when these are withdrawn at night by the decreased radiation from the earth, the cloud sinks and disappears. Hence it is that when radiation begins in the morning with the warming rays of the sun, clouds are formed, and when it ceases at evening the lower clouds disappear and only the high and comparatively dry ones remain. *Day and night clouds.*

The meteorologists have established four broad classes of clouds according to their different forms, and the different heights at which they are usually seen. The classification is largely for the sake of convenience because, as has been already intimated, clouds are substantially the same thing whether high or low in the air; and the different forms run into each other so closely that it is often difficult to tell one from another. The four classes, beginning with the highest and ending with the lowest, are the cirrus, the stratus, the cumulus, and the nimbus. There are some subdivisions which may be recited in order, but the broad divisions are given at first to avoid confusion. *Classification of clouds.*

THE CIRRUS (1) is the frailest and the lightest of all the cloud forms, and drifts at the greatest altitude. It is sometimes seen fifty thousand feet or more above the earth, though its usual elevation is not so great. Apparently it stands still in thin wisps and curls against the blue, but in reality it is a rapid traveller with the wind, and sometimes reaches so great a velocity as ninety miles an hour. It is not a large cloud and in form is curled like hair, is fibrous, or perhaps feathery. At times it is streaked across the sky in a light film somewhat like the Milky Way, but more frequently it is in small, thin patches. It has also many patterns that resemble stripes, tails, plumes, and wings, but they are all diaphanous and film-like. When it appears in streaks and lines these are usually parallel to the wind, and are commonly spoken of as "mares'-tails," "goats'-hair," or "cats'-tails." These clouds often move in irregular, straggling groups. There may be only a few straw-like wisps, and then again the upper space may be spattered with them. Too thin and nebulous as a general thing to show shadows, they are the brightest of all the receivers and reflectors of light. This may be for two reasons. First, they are higher than any other clouds and receive a more powerful light from the sun because of the clearness and thinness of the air in which they drift; secondly, they are ice-clouds, that is, made up of needles of ice, and are more capable of reflecting light than the ordinary vapor clouds. Certainly their luminosity is their strongest feature aside from their peculiar spray-like or feathery form, though their color is often remarkable. At dawn they are the first ones to catch the light from below and reflect it in yellow or pink, and at twilight they are the last ones to fling back the scarlets of the sinking sun. These clouds are apparent in all countries and in all skies, and are ever tenants of the upper region, though some of their branches or manifestations appear in connection with clouds of the middle region. *Color of the cirrus.*

The cirro-stratus (a) is a mixed or composite cloud made up from the cirrus and the stratus. It is not one of the four large classes, but rather a hybrid variety that must figure under a subdivision. In reality it is a part of the cirrus, which has become slightly changed in its form and elevation by a sudden increase in its moisture. Grown heavier and denser, it has descended and woven itself into long, thread-like lines resembling a net or veil stretched across the sky. Its appearance is usually thought to be indicative of approaching storm, and the direction it takes shows whence the storm is coming. It is a frost cloud, is frequently seen a *The cirro-stratus.*

an altitude of thirty thousand feet, and has a maximum travelling velocity of about seventy miles an hour. It is the substance from which the halos about the sun and moon are woven, and is very thin, almost transparent. Like the cirrus, it casts no patches of shadow, is pale white, and when struck from beneath by the rays of the sun below the horizon is marvellous in its delicacy of light and color.

The cirro-cumulus (b) is another mixed cloud. When the cirrus descends still lower than the region of the cirro-stratus, its edges of frost begin to melt like the sharp sides of a snow-bank. It then takes on a woolly appearance similar at times to the small, detached portions of true cumulus, though it lies in a much higher field of air. It has a fashion of breaking up into small, rounded patches like rotten ice in a river, and of drifting across the sky in vast companies that almost hide the blue. There are two forms in which it appears. One is called the "dappled sky" or sometimes "wool-pack" from its fleecy nature; the other is the "mackerel sky," which is not fleecy but hard-looking. The latter is rarely seen as compared with other cloud forms, and in England it is always thought to be the

Sun and moon halos.

The cirro-cumulus.

Dappled and mackerel skies.

harbinger of fair weather. Both forms of this cirro-cumulus are frost clouds. They drift at an altitude of about twenty-two thousand feet, and have a maximum velocity of about eighty miles an hour. Their movements across the sky seem to be systematic and orderly, though of course the regularity of their driftings is dependent entirely upon the steadiness of the upper wind-currents.

THE STRATUS (2) is a flat sheet cloud extending in long lines across the sky, at times bridging it, covering it from horizon to horizon. It is the cloud, let us say, of the middle-air region, though every cloud that has a sheet-like form or looks stratified is some kind of stratus. It is usually formed when there is little wind and only a mild radiation is going on. The air as it rises gets gradually cooler until the dew-point is reached, when this cloud forms and extends itself across the sky in long, thin drifts like the smoke from factory chimneys in calm weather. In color it is a gray cloud, though occasionally, when very thin and the sun or moon is shining through it, it looks bluish in tint. At times it has a concave or a convex appearance, and at other times it is wavy or undulating. It is from ten to twenty thousand

The stratus.

feet above the earth, and though its movement is hardly perceptible to the eye, it may be drifting at the rate of fifty miles an hour. Its effect in making a hazy day is quite noticeable, and at sunset, when it lies across the western horizon in bars, it is often very pronounced in reds or chrome-yellows.

The strato-cumulus (a) is another and perhaps more common form of the stratus. It is a heavier variety, darker in color, and more roll-like in form, caused by its having about it something of the lumpy nature of the cumulus, yet with enough of the stratus to make it form in a layer along the sky. It is a cloud that may send forth rain, though it often overhangs the earth in dark folds for days at a time without giving forth a drop. At times it looks like a compact, dense rain cloud, and when it assumes this shape it is often confused with the nimbus.

CUMULUS (3) is the name given to any cloud that has a heaped-up, mountainous, or lumpy look about it. The white patches that bowl across the sky on a summer's day are detached portions of cumulus; but the most noticeable form of it is the "heap" cloud that on warm afternoons lies off in the southern sky, rising

The strato-cumulus.

The cumulus.

far upward toward the blue in fantastic turrets, domes, and peaks. The bases of these clouds are usually dark in shadow, flat, and cut sharp; while their tops are cast in wreaths and billows of vapor. They appear at times to be of great height, for though their bases are usually not more than five thousand feet up, their tops sometimes reach forty thousand feet from the ground. At such an altitude the crests look woolly, which probably indicates that the cloud has reached a cold region and has changed to frost-dust on its top. Usually these clouds appear to stand firmly and to be motionless, though they are always changing, their bottoms sinking away and their tops being continually renewed. Moreover, they are drifted by winds at the rate of about twenty-five miles an hour, though at other times they may scarcely move at all. After sunset they usually sink and disappear entirely. *Cumulus changes.*

The heavy cumuli are summer clouds, and are not seen in cold climates nor upon cold days. The tropical region is their home, though they are native to the temperate zones in midsummer, and are often seen rising along the horizon like a range of snow-clad mountains, with hills and valleys running up or down *Summer clouds.*

or across them. In outline they are graceful, and in light-and-shade they are often sharp-marked. The best time to study them is in the evening, when they are lying back at the south or east. Then the pinnacles and peaks glow with light, and make the snowy-mountain illusion more palpable than ever; or they turn into phantom, rock-based promontories with spectral tides of vapor at their feet that sound not and shock not, yet rise slowly higher and higher upon the snowy walls. Occasionally a tall, heavy mass is veiled by a thin layer of the stratus, through which the form of the cumulus is seen to burn like a great opal. Sometimes, too, a heavy cumulus is seen through city smoke at sunset glowing like molten metal. When in the west and in front of the sun this cloud is the one that shows us the gold or silver lining; and under sunset light it is possible for it to take on all tints and shades. When it is not near the sun but lies off at the side, we often see the pink, "Alpine glow" suffusing the white castellated tops; and the shadows caused by sharp breaks of form often show blue, lilac, and even pale green in hue.

The cumulo-nimbus (a) is substantially the same cloud as the cumulus except that it drifts at a slightly lower level, is not a tall tower cloud, and has in it an admixture of the nimbus or rain cloud. It is in fact a form of rain cloud and is responsible for the "sun-shower" as well as for others of greater force, like the thunder-shower. It is also a cloud that shows a silver lining when seen against the sun, and at night it reflects heat-lightning very brilliantly. In the daytime its base appears dark, its top light; and at twilight, when lying off in the east, it banks up at times like a table mountain in layers and terraces that reflect the pinks and violets of the sunset. Its usual altitude is about four thousand feet, and its movement is more rapid than that of the cumulus. *The cumulo-nimbus.*

THE NIMBUS (4) is the rain cloud, and every cloud from which rain falls is some form or combination of the nimbus, though the nimbus proper is the flat, sheet-like or rolled rain cloud. It is the closest to the earth of all the clouds and is consequently the first one to receive the smoke, dust, and heat arising from the earth. By comparison it is a foul cloud, and is for that reason a rain cloud—the formation of vapor spherules being, perhaps, dependent upon the presence of dust-particles in the air. The nimbus takes all forms according to its density and *The nimbus.*

velocity. In afternoon showers it resembles the cumulus; in driving storms it lies lower to the earth, moves in great, rolling puffs, or flattens out into thin, fast-flying sheets with ragged edges and long, projecting arms like antennæ. At times, when a storm is prolonged for days, the forms of the clouds are hardly discernible; the masses are lying low in the air and spread from one to another with such close connection that they look like one vast stretch of gray across the sky. In thunder-storms these clouds often bank up dark and threatening in the form of an advance-guard. They move forward quite rapidly and carry with them a rushing wind. The first-comers are always the darkest-looking and most violent of the storm, yet they give forth neither lightning nor rain. They seem to be only wind-makers, though it is common knowledge that clouds are not makers of wind, but merely manifestations of wind existent. The gray clouds behind the dark advance-guard are the ones that carry the rain. In tornadoes the darker ones often twist, writhe, and roll over one another as though pulled by a violent under-current of wind; in cyclones the movement is similar, but from an opposite cause. In the latter case the pull is likely to change to a push caused by rising swirls of heated air trying to escape up a vortex into cooler regions. The color of the nimbus is always cast in gray, and the darkest portions are usually the ones under deepest shadow. Poet and romancer to the contrary, there is no such thing as a "black" cloud seen in the daytime—nor for that matter at any other time. The heavy storm-cloud may border upon purple, and sometimes preceding cyclones it is sea-green, but it is never "black."

The different forms and kinds of clouds given above enumerate only certain families. Aside from the large groups there are patches of cloud being continually woven or torn from one family to blend and intermingle with another family, thus making many hybrid varieties. It would be almost impossible to catalogue the different cloud forms that one may see on an ordinary summer day; or the parts of clouds such as scud, wrack, wreaths, and sprays wrenched away from the parent body by storms and squalls.

Scattering cloud forms.

The form of clouds usually gives the ear-mark of recognition to such families as the cirrus, the stratus, and the cumulus; and yet this form is never the same for any length of time. It is always shifting, changing—being added to or subtracted from by varying conditions. It may describe the species, and yet is hardly to be called the characteristic feature. That which strikes us as peculiar and determinate about any and every cloud is its drifting, swaying lightness. The glide down a vast incline of air that marks a white swan settling to water is usually considered the most poetic of all motions; yet it is somewhat gross and heavy compared with the grace of a moving cloud. A cloud drifts *with* the wind, not before it; it lies *in* the air, not beyond it; it has no visible support and yet appears supported. Apparently defying the law of gravitation, it seems to have no relation to the earth, but like a phantom ship sails the celestial blue, wholly unconcerned as to destination, wholly careless as to dangers. All of them, singly or in flocks, are mere vapors — such things as dreams are made of—the wonder-world of childish fancy, yet how beautiful they are!

The lightness of clouds.

Cloud fancies.

> "Forming and breaking in the sky,
> I fancy all shapes are there;
> Temple, mountain, monument, spire,
> Ships rigged out with sails of fire
> And blown by the evening air."

They rise, fall, or change before our eyes with no effort, no sound, no apparent design. Now they are scattered wide over the blue, now they are huddled together and driven in flocks by the wind; but they never seem to be in a hurry. An epitome of idle content, having no actual power in themselves, they are, nevertheless, the visible sign of aërial energy. The wind blows them whither it listeth. They drift around and about the world and have no abiding-place, no resting-place on land or sea; yet wherever they go they gladden the eye and cheer the heart, and in every landscape they are the bright spots of beauty. *Drift of clouds.*

And what wonderful luminosity there may be in a cloud! The upper cirrus just before sunset is often dazzling in its light, and when struck full by the sun's rays, there is nothing more intense in luminosity than the cap of the tall cumulus. The ancients felt the splendor of this cloud light, and it is not strange that the Old Testament writers should speak of the "pillar of cloud" that guided the wanderings in the wilderness, of God descending on a cloud, of a cloud as the resting-place of the Mercy Seat and the standing-place of angels. The purity of these white vapors of the upper air *Light upon the clouds.*

seems uncontaminated by earthly touch, and their shining surfaces are not comparable to any terrestrial thing save the newly fallen snow glistening on the highest Alpine peaks.

Cloud colors.

And in color what is, what could be, more gorgeous, without a note of discord, than the western clouds at sunset? They have no hue in themselves, and yet, like the flowers of the fields and the waters in the lakes, they have the power of reflecting and refracting colors of the utmost brilliancy. And how vivid these hues become as the hot sun throws his parting shafts of fire over and under and through the fleecy drifts of vapor! After the red disk has fallen below the horizon the scattered patches continue to burn and glow in scarlets, golds, and pinks—all imaginable hues from bright blood-red to dark violet. As the sun sinks still lower its shafts strike upward upon the under-surfaces of the clouds, and for a time the color seems even more brilliant. And when the

Clouds at sunset.

cloud-bars just across the horizon begin to dim their lustre the high, "mackerel sky" catches up the color and the flame mounts upward to the zenith, from cloud to cloud, like steps in a ladder of fire, lessening in glory as the height is reached, and finally lost entirely in the blue.

Last of the twilight glories, when the light has gone out of the lower clouds and the white cumulus has turned to dark purple, the wavy forms of the cirrus may be seen flaming like wind-blown torches far up the western sky.

Common as the sunset colors are, we never seem to weary of them. They are always things to look at and to wonder over. No hues seen upon the earth are so full of light and fire, so brilliant in variety. The colors of the rainbow show a celestial spectrum, but they seem to pale beside the sky-splendors of the west; and as for the colors of the clouds at dawn, they are much paler than those of the sunset. At noontime the clouds show no color in particular. Occasionally low-lying cloud flocks over a city like London will have a heated, flushed look, and when close to or under the sun they will glow like plates of hot iron; but this is caused by local dust and soot in the air. Often, too, in all warm countries a cloud passing across the face of the sun will have silvery or golden edges, and a pyramid of cumulus may be pink in the lights and blue in the shadows; but, aside from such exceptions, the clouds at noonday are practically white in light or grayed under shadow.

Clouds at dawn and noon-time.

Clouds in landscape.

We realize quickly enough how important to our enjoyment of landscape are the sky and the white clouds as soon as they are cut off from our view by the drawn veil of a rainy day. The variety of color in the sky and of movement and form in the cloud, the feeling of space, distance, loftiness in them both, are gone; and with them perhaps the most effective features of all landscape. Anything that obscures or shuts out sky-space, with its interminable depths of blue and its bright clouds, mars one of nature's greatest beauties. Even a horizon-line so high as to narrow the sweep is objectionable; and hence the valleys of the Alps, though grand enough in view of mountain bulk and snowy peak, are the least livable places in Europe. The great palisade of rock breaks the reach of the sky and we lose directly in color, light, and atmospheric perspective. On the contrary, a flat, low-lying land, though perhaps the last to

Sky lines in landscape.

be loved by humanity, is in the end the most livable of all. The prairies of North America, the plains of Lombardy, the flat lands of eastern England, are supreme in the feeling of space in sky and of distance in cloud. Something of the great charm of Venice lies in her flat lagoons and her great, uplifted sky; and to those who know their book of landscape well, the green fields of Holland arched by blue and white are the most restful, enjoyable, serenely beautiful lands on the face of Europe.

CHAPTER V

RAIN AND SNOW

IN order to understand the phenomena of rain and snow we must consider for a moment some facts established by the weather men. I have no notion of trenching upon the domain of the meteorologist. Indeed, I had thought to write a book that would suggest some of nature's beauties rather than its bare facts, but I find it continually necessary to explain beauty by first showing structural character.

The vapor-carrying capacity of air.

The capacity of air for receiving and holding vapor depends upon temperature. It is small at low temperatures; it is large at high temperatures. That is to say, the vapor-carrying capacity of a cubic foot of air is ten times as large at 100° Fahrenheit as at 32°. At either temperature, when the cubic foot has all the vapor it can carry, it is called "saturated." When more vapor is crowded in than the cubic foot can carry the result is condensation of the surplus into cloud and rain. Perhaps this can be illustrated in a simple way by putting a suppositious case in which I shall use the figures of Dr. Robert Mann.

At 32° Fahrenheit a cubic foot of air can hold or carry 2.37 grains of vapor in invisible form. It is then said to have reached its "dew-point." If into that cubic foot of air 2.38 grains of vapor were injected, the result would be one-hundredth of a grain of condensed mist or cloud. At a temperature of 60° each cubic foot of air can carry 5.87 grains of invisible vapor; at 80° each cubic foot can carry 10.81 grains. Consequently, if at any time or for any reason a saturated air at a temperature of 80° were suddenly chilled down to 60°, nearly 5 grains of surplus vapor would be condensed out of each aërial cubic foot in the form of tiny droplets of rain.

The dew-point and condensation.

If at a temperature of 32°, the freezing point, similar conditions prevailed—that is, if a saturated air at 32° were suddenly chilled down to zero—a similar surplus quantity of vapor would be condensed in the form of crystallized spicules of ice or snow. A more violent reduction in the temperature of a saturated cloud—say from 100° down to 60°—would produce more vapor than the cloud could hold, and it would inevitably fall to earth as a shower.

Causes of clouds and rain.

It is now generally held, I believe, that the cause of clouds and rain is largely, if not entirely, the cooling of air by expansion as it ascends; and that intermingled cold and warm air, and the chilling of air by cold bodies such as mountain-tops, have little or no effect. Certainly the expansion of air is the final but not always the most immediate cause. The chilling produced by warm air driven against cold air and its result may be frequently witnessed in the winter season along the Atlantic coast of North America. When the wind shifts to the east we are all quite sure that thirty-six hours at least will bring rain, and usually it is not so long before the clouds begin to drift inland from the sea.

Eastern storms, how produced.

It is sometimes thought that there is a storm on the ocean, and that it has been travelling landward for hundreds of miles. Occasionally that is the case, but more often the clouds and rain are formed along our own coast, and in this way : The sea is much warmer than the land, especially in the Gulf Stream region. Vast bodies of moist air overhanging it are driven in upon the land by the eastern winds. This land is ice-locked and very cold.

As soon as the warm air of the sea meets the cold air of the land a chilling-down process begins and condensation into clouds is the result. The coast is the line of condensation, and as these clouds move into the cold interior their vapor-carrying capacity grows less and less until finally rain is precipitated.

Another illustration of cloud and rain making is often seen in the spring of the year, when a warm air blowing from the south meets a cold air blowing from the west. The warm air is forced up and over the cold air, clouds are formed all along the line of contact, and heavy rain is not the unusual result. Again, a sirocco blowing up from the south across the Adriatic will make the cool stones in the pavement of the Piazza San Marco at Venice "sweat;" and when this sirocco meets the Southern Alps and is tilted up into the cold snow regions of the peaks, condensation, clouds, and rain follow. *Warm winds and cold mountains.*

Just how the rain-drop is formed seems not better known than the constitution of the spherule of moisture in the cloud. A recently advanced theory would seem to argue that moisture forms upon and about the tiny dust-particle in the air, using the particle as a nu- *The rain-drop.*

cleus, so to speak, and that by augmented condensation the spherule gradually grows to a rain-drop. Once formed, the drop has about it an elastic skin or envelope that prevents it from breaking unless pressed or struck by some body. Oftentimes it preserves its form against sharp shocks, as we may test by shaking the dewdrops on flowers, or observing the drops from a fountain that run across the surface of the water like pearls for some distance before coalescing with the main body. In the air the rain-drop is always perfectly round, as the camera shows us, even if it were not a necessity of that phenomenon, the rainbow.

Its elastic skin.

The size of the drop is doubtless dependent upon the amount of surplus moisture in the cloud. This in turn is dependent upon the temperature of the air and the extent to which this temperature has been reduced. Doubtless, too, the suddenness of condensation has something to do with the size; and besides that the drop in falling probably unites with other drops, somewhat as globules of mercury coalesce, or a rain-drop running down a window-pane gathers other drops in its downward course. That the temperature has much to do with the quantity of vapor in the cloud, and

The size of the drop.

consequently the size of the drop, we may believe when we consider how small are the raindrops in winter and how large they are in summer. The first ones falling in a thunder-shower, for instance, are unusually large. Possibly the size is caused by the outer edge of a heavily saturated cloud being driven by the wind against cold air and swift condensation following the meeting ; or it may be that the heavy drops fall from a very high cloud and coalesce with others in falling. It is usually only the first-coming clouds that cast the heavy drops, and after the first dash they grow finer, smaller, and more numerous. *The first heavy rainfall.*

A thunder-storm comes and goes quickly, the moisture being in measure localized. Both its coming and its going present interesting, sometimes fantastic, forms of clouds that are continually torn, scattered and reunited by the drive forward of the wind. Usually the cloud is a thick one, and in its lowest part is dark, becoming lighter in its main body, and if it is a towering, cumulus cloud, its upper peaks may sometimes be seen before or after the storm, shining white in the sunlight. Beautiful by day, all the forms of thunder-clouds are even more beautiful by night, when lightning flashes *Thunderstorms.*

illuminate them. Then they have a pale-bluish coloring, the light-and-shade upon them is clear-cut, and the feeling of massive form is convincingly brought home to us. The great, dark clouds lying underneath seem but the flat pedestals of the white peaks and spurs that far up the zenith seem to tower and rock slowly like icebergs on a stormy sea. At other times the clouds seem softer and roll upward in billows and wreaths—great vapory masses of blue-white that boil and seethe with the force of the winds. And how the currents of lightning pass through these heavy clouds without producing the slightest disturbing effect upon them! If lightning were shaped like the classic bolt of Zeus, or zig-zagged and raw-edged, as popularly depicted, it might disrupt even cloud forms; but instead of that it runs in streams and rivulets, and when seen in photograph it often looks like an outlined map of the Nile, with its many mouths leading to the Mediterranean.

Lightning and clouds at night.

Another accompaniment of the thunder-shower is the fringe of rain that may be seen trailing from the clouds as the shower passes to one side of us. This fringe waves slightly with the wind, and when seen at a distance looks as

Rain-fringes.

though it did not reach to the ground. As a matter of fact, some precipitations never do fall to earth. They are evaporated in mid-air and returned to the sky. The travel of the rain-fringe across the country, veiling and often obscuring the hills and meadows, is most interesting to watch as it shifts its form, color, and density, and darkens the green of the country over which it passes. It changes more frequently than we think, and is sometimes temporarily lost before our eyes, only to reappear again with startling brilliancy when struck by a chance sun-shaft. When the shower comes our way, the clouds themselves seem to undergo changes as soon as the rain begins to fall from them. The lumpy roll breaks and flattens in strata, or else it trails down in long, shaggy points. The whole landscape darkens as the shower approaches, the clouds become obscured, the trees blurred, and presently we are in the centre of a circle of rain through which we can perhaps see not more than a few hundred feet. When the shower is passing away, everything is, of course, reversed. The light increases, and often the vanishing rain clouds struck by the sun, gleam as frost-white as the castle-clouds of a summer afternoon.

Travel of the rain-fringe.

Circled by the rain.

The rainbow.

With the sun shining after a thunder-storm, and the light striking upon the clouds beyond us, comes one of the most noticeable beauties of the sky, the rainbow. It is caused by the drops of water in the air becoming prisms of light and casting the spectrum colors. A thin sheet of these falling drops is struck obliquely by the sun's rays, and each drop has light entering the upper portion of it, and undergoing two refractions and one reflection. The exact scientific explanation of the arch of light, and how it casts the colors of the spectrum, is foreign to the present purpose. Suffice it to say that the arch is seen only when sunlight strikes falling rain obliquely, and that it shows the colors of the spectrum, beginning with red on the outside. The secondary or upper bow is like the first, only fainter, owing to a double reflection within the drops, and with the colors reversed—that is, the violet is on the outside.

The moonrow.

The bow caused by the moon is much fainter than that caused by the sun, and is not frequently seen. It rarely shows distinct colors, and is most commonly seen as a pale gleam of white or yellow light.

The three-days' storm of rain, common to all temperate climates, is quite a different affair from the thunder-storm. It begins with no such frowning front, but has infinitely more endurance because it is not localized. The clouds are spread over a large area of sky and they gather themselves together slowly at first. When condensation sets in and rain begins to fall it is slight, almost like a Scotch mist. But it soon gains in power, the wind rises, and the small rain-drops begin to drive toward the earth with great swiftness and force. The heavy drops of the thunder-shower, falling a long distance from high clouds, and falling straight, seem to have much less striking power than the smaller drops driven diagonally by the wind. Nor is the wave of a rain-fringe from a thunder-shower anything like so violent as the sheet of driving rain in the three days' storm. The latter shakes banner-like in the wind as though it were a veritable sheet held down from above, or it rolls in swift-moving undulations across the sky like the wavy light-flashes of the aurora. *Three days' storms.*

But there is little in the long storm to be admired or enjoyed unless we ourselves happen to be in a tempestuous mood. The domed sky is shut out, the clouds make a flat, lead-colored roof overhead, or else they form in gray billows *Rainy days.*

like an inverted sea in storm. Color is gone save the vast monotone of gray, and form is almost obliterated except in the lines of falling rain. The splash and beat of gusts upon the roof and the window-pane, the moaning and raving of the wind, are rather dreary; and without, everything is even more dismal. Decidedly the best place is by an open fire with a book in one's hand. When, however, the rain has passed, and the sun is once more seen, we have an irrepressible desire to come out from hiding, like the birds, and see what the rain has done for the world about us. The freshness of nature, the smell of the ground, the clearness of the air, the brightness of the vegetation—the feeling as though the earth had had a bath and was waking, clean and refreshed—are omnipresent. Color, too, seems revivified. The geranium and the rose are more brilliant, the grass greener, the trees more luminous, and overhead the blue sky is deeper in its coloring and light than possibly we have ever noticed before.

After the storm.

This is all more marked in the country than in the city. The only noticeable thing about rain in the city is that it washes down the buildings and cleans up the streets. The patches of grass and the trees in the parks do not seem to respond to it so quickly as those on the lawns and fields out of town. This may be imagination with the observer, and yet it is well known that the rain which falls in the city is not the same rain as that which falls in the country, though both precipitations may come from the one cloud. City rain is fouled by passing through smoke, dust, and gases. It gathers sulphuric acid, which corrodes metal, paint, and iron, and certainly does not help vegetation. The country rain is always purer because falling through a clearer air. *City vs. country rain.*

Precipitation from the clouds usually takes the form of rain and hail in the summer, sleet in the spring, and snow or frozen ice-crystals in the winter. They are all easy to account for as regards their forms except hail, which is frozen rain perhaps, but a satisfactory explanation of how it is formed and frozen has not yet been offered. Hail falls in hot, sultry weather and with a thunder-storm. For that reason it is suspected that it has to do with electricity or is caused by it. It would seem at first blush as though those heavy drops of rain, which have been spoken of as the first to fall from the thunder-cloud, were sometimes congealed to ice and united to other drops in the congealing proc- *Hail.*

ess, and that hail was made in that way. The two precipitations, one in rain and one in hail, correspond in time, place, and circumstance, and apparently are identical with one another; but the perplexing question arises, How does hail freeze in its peculiar form? If a rain-drop falling from a warm cloud should pass through a very cold current on its way earthward, it would be frozen into transparent ice; but that is not the make-up of the hail-stone. The centre of the stone is opaque, milky, cloudy, as though it were a tiny, frozen snow-ball; and around this centre are usually thin, concentric layers of ice and snow formed like the layers of an onion. From its appearance one might say that it was a frozen particle whirled around through rain and ice clouds, gathering bulk to itself by contact, much like a snow-ball rolling down hill on a moist, winter day.

Formation of hail.

The theory has been advanced that the rain-drop is caught up by powerful, ascending currents and carried to regions of snow and cold, and afterward allowed by the declining winds to fall back to earth; but if so, how does it arrange to get back in time to form the first fall from a thunder cloud? It is more probable perhaps that the top of the thunder cloud reaches up

Hail theories.

into the snow regions of the air, and that precipitation falling from it in the shape of snow gathers bulk to itself in descending until, passing through the rain region, it adds an outer coat of ice. The hail-stone certainly falls a long distance, as we may know from its striking power, but whence it falls, and just how it is formed, the meteorologists have not yet definitely told us.

The hail-stone is usually not larger than a cherry, though in description it is sometimes "as large as a hen's egg;" and it has been seen as large as a good-sized apple, but not in the temperate zones. It is elastic, and the bounce of hail from the walk or lawn is a commonly observed fact. Sometimes with wind it drives diagonally to the earth, but more frequently it falls like the heavy drops of the thunder-shower. Usually there is nothing marked about its color. It is lighter in tone than rain, and when falling through the air shows blue-white. At times a very beautiful effect is produced during sun-showers by the sun's rays flashing upon the stones as they fall. They are then dazzling opal-white, and quite different from the rain-drops, which fall through sunlight like glittering diamonds. Occasionally one may see a hail-storm turned into some-

Hail-stones.

thing like a rain of fiery red or yellow pebbles, by having the shower between him and a red or yellow sunset; but this effect is of rare observance.

Snow is the excess vapor in the air condensed into spicules of ice. It forms whenever the temperature is below freezing, and many are the forms of flakes produced by the crystallizing process. *Snow flakes.* When the fall is light and feathery, owing to a low temperature, countless variations of the six-pointed star may be seen on a dark ground, such as a coat-sleeve. When the temperature is higher there is a tendency toward agglomeration, or the union of many flakes into one. Snow falling from a cold into a warm stratum of air is softened around the edges, and we have what is called a "wet" snow—that is, a snow containing considerable moisture and in form large and fluffy. The reverse of this takes place when the snow is falling from a cloud warmer than the temperature of the lower air. Then we have a hard, round *Different forms.* snow, sometimes called "ball" snow. It would seem to be hardened and compacted by passing through the colder, lower air; and when it reaches the earth its form is that of the fine snow that falls in the long, cold storms of winter.

In the highest clouds snow is always a possible and often a necessary result of condensation. When it falls it frequently melts into rain in passing through the warmer and lower air. The storm that covers the top of Mt. Blanc with snow falls as rain in the valley of Chamonix. Many of the high mountains have their snow-line above which rain is not known, and we hear their peaks spoken of as being covered with "eternal snow." The words are not accurate, to be sure, for snow even on mountain-tops is continually melting and passing away into glaciers to be replenished by new falls; but the description is true enough in the sense that the peaks are always snow-capped. *Snow on th mountains.*

At the beginning of a snow-storm the flakes are few and large, and they settle to the earth like eider-down or thistle-spray. Nothing can exceed the gentleness of these first-falling flakes. They whirl and float and hover and fall so softly, that not a leaf or grass-blade is stirred; and they melt into the smooth surface of the lake without making the slightest visible impression. And how absolute the silence of their fall! One by one they gather together on the earth without a sound, and in the morning when the children look out of the window they are sur- *The first fall.*

Snow-storms.

prised to see the white fairy-land, and they had no intimation whatever of its making. During the day, if the storm increases, the flakes are likely to grow smaller and harder—the fall being much like the smaller rain that follows the few large premonitory drops. With a high wind the snow drives almost horizontally at times, and when the wind is in gusts the snow-sheet waves more lightly and easily than the corresponding rain-sheet.

The blizzard.

In Northern countries the light snow driven by high gales often results in what is called a "blizzard"—something almost impossible in the region of New York, though the name has been and is frequently applied to every severe snow-storm. A blizzard proper, such as they have occasionally in Dakota, brings with it a fine, driving snow that strikes the face like a shower of sand, stinging, cutting, and almost blinding one. The temperature during its prevalence is usually so low that there is little or no moisture in the air, and the blowing of the wind does not allow the snow to catch and lie upon the ground except in sheltered places. Gusts and eddies are continually swirling great sheets of it through the air. If the ground was previously covered with snow, the low temperature has possibly prevented it from having anything like a crust upon it, and the first sweep of wind raises its light particles in the air to join the new-comers. The total result is blinding and confusing to the wayfarer. The air is full of flashing, dashing flakes, and one can see no farther in the maze than in a dense fog—often not so far. All landmarks, roadways, and trails are obscured in a few minutes, and people perish in such storms through losing their way and being overcome by the cold, the wind, and the driving snow.

Once fallen, a mantle of snow produces the most decided change in the appearance of the earth, excepting the change from night to day, of which we have knowledge. The earth is naturally a light-absorber. It drinks in sunlight and reflects just as little as possible, so that its general appearance is comparatively dark, with sheets of water showing here and there as spots of white. When snow covers the ground the appearance is reversed, and such objects as trees and bare rocks appear merely as spots and patches of dark upon the white. The intensity of this white is common knowledge. It is a bluish-white and much lighter than the clouds casting it forth. This is largely for a reason

The white cover.

already given. That is, to repeat it, looking up we see the shadows of countless cloud-particles; looking down we see light reflected from countless snow-surfaces.

But the intensity of the white is not wholly explained by the difference between reflected and shadowed light. There is another reason for its whiteness, and perhaps it is not uninteresting to know that if the new, fine snow is examined under a magnifying glass each separate flake will be found to disperse as well as to reflect light, and everyone of them will show prismatic edges casting the rainbow colors. These colors are the component parts of light—light disintegrated, in fact. The tiny prisms scatter the light into colors, but the mass of them taken together reunite the colors into light. It has long been known in painting that small stipplings of red, yellow, and blue, placed close together, will throw out more light than a pure white ground. Light recomposed from colors is stronger than light reflected. It is this principle, practically demonstrated by nature, that lends something of peculiar brilliancy to the newly fallen snow. And how brilliant, how dazzling is that newly fallen snow only those know who have seen it in very cold coun-

Snow prisms.

tries, where vapor is a practical impossibility and only the ice- or snow-crystal exists. In such lands the covering of the earth glitters as though thickly sprinkled with diamond dust, and the mist rising from swift-running streams is frozen into hoar-frost that drifts in the air, sparkling in the sharp sun-light. It is flash and gleam from every point of view as though a dozen suns were in the sky and all were flaming brightly.

Brilliancy of snow.

This splendor is greatly modified in the regions where the snow is moist and forms in heavy masses, loading the branches of the pine and the spruce, muffling the eaves and chimneys of the houses, and piling up in pyramids on the tops of the gate-posts. The brilliancy is pronounced for only a few hours. Under the sun and its warmth the crystals lose their sharp angles and melt down into ice-particles, the pyramids soon slip from the gate-posts, and the pine, shaking its long branches in the breeze, throws its burden of snow from it. The purity and serenity of the morning following such a snow-fall, when the sun is up and we are out walking the fields and woods through the still whiteness, are not lost upon us. We all feel the solemn beauty of the scene, the hush

The snowy landscape.

of the earth, the dark ranks of trees, the gleam of the cold sky, the glitter of the snow lying so fluffily upon earth and tree and hill and house-top. How calm and pure it seems! How impressive it is, too, under moonlight, with the hills stretching far away in their white, heaving mantle, the frozen woods standing up so darkly along the night horizon, the stars glistening in their violet depths, and over all the great silence of the sky!

Under moonlight.

And what a multitude of sharp angles, harsh forms, and bleak colors are hidden under the muffling of snow! The ragged mound, the rough cornfield, the tumbled meadow, the bushy foot-hills of the mountains are smoothed out, and evened over, and cast in new forms. Everywhere there are flowing, rounded lines running hither and thither to meet other lines, intertwining and uniting in graceful and rhythmic combinations. In the open fields, where the wind has been at work, the snow may be cast in rolls, like the long swells of a smooth sea; and when the sun is low these swells show pink light on their crests and blue shadows in their hollows—shadows even more delicate and tender in hue than those cast upon water. Above the open fields even the mountain-lines

Snow lines.

Snow colors.

against the sky are softened by the snow; and the ragged promontories, smoothed into heaving mounds of white, glow with a pinkish hue under the sunlight and at evening turn to cold purple.

And how sharp is the contrast where the river runs darkly flashing through banks of snow that come down and meet the water's edge! It is a picture in black-and-white. The bend and sweep of the lines in the banks are clear-cut and sharp, defining on either side the flow in and out of the most graceful thing in the world—running water. There is nothing more rhythmical than the curves made by water, and the flowing river in winter is emphasized and intensified by its white borders. Sometimes it happens that the stream is frozen with clear ice, and then from a high point like a bridge, when the wind is blowing, one may see little rivulets and streams of snow running over the top of the ice, following channels, swirling and eddying almost like the stream itself except that the motion is much faster and more serpentine. Very graceful are these little currents of snow. They may be seen again chasing, whirling, and drifting on the crusted and frozen fields, but not so readily as upon a dark background of

The river through snow.

Swirls and drifts.

ice. The winding courses they follow and the beautiful forms of snow-drifts into which they finally resolve themselves, are distinct features of the snow-landscape.

In early spring.

The early days of March when the snow is beginning to melt, when the rocks on the hillside heave out of the white, and odd patches of ground show dull gray or brown, are usually considered the dreary days of the year. Most people declare the country "stupid" at this time and house themselves in cities if they can; but to some nature-lovers it is perhaps the most interesting season of all. The snow on the side-hill still lingers; but the meadows are bare, the brooks are swollen, the ice is gorged in the river, the valleys are shining with pools of water.

The skeleton of nature.

The skeleton of nature is pushing through its winter mantle at every point; but if we look at it with appreciative eyes we shall find the hills and the rocks and the bare trees beautiful as outlines merely—beautiful in their rugged, broken angles and their traceries of line against the snow or sky. Besides, there is some little color noticeable all through the winter in the red stems of the maples and the birch, in the ruddy glow of the swamp bushes. This color begins to heighten in March and with it comes the sense or feeling of stirring life. It is in the very air. Nature is turning as though anxious to rouse from slumber. The evidence of life is not great, but we feel under stillness, coldness, and bareness a potential power. The great oaks and chestnuts that stand high up on the mountain, their trunks showing against snow-banks, their branches against the sky, will soon be turning green, and the meadows and swales of the valley will glow with new life and color.

Stirring life.

Perhaps just at this time, when nature has not yet started out of winter, there comes a late snow-storm which turns to rain, covering the limbs of the trees with ice and putting a crystal coating upon the earth. Then what a spectacle we see the next morning, with all the world glittering like spun glass under the rays of the sun! It is a brilliant sight, and at times a most astonishing one in color. For, if we can get the ice-bound trees between us and the sun they will take on any color that the sun or sky may show. Occasionally, with a red sunset, a whole grove of trees will look to be on fire, and under a yellow sunset the same grove of trees will appear of the most brilliant topaz hue. It is not unlike a similar effect seen in falling hail.

Ice-locked branches.

The awakening of nature.

The icy landscape is not a sight of any long duration. The sun soon melts the ice, the trees rock in the wind, and the glassy covering slips and rattles upon the frozen ground. When once nature begins to move, it is not easy for cold winds and blustering sleet to stop it. The grass starts under the snow, the early plants begin to stir, the stems and buds grow redder; and when the last patch of dirty white in the deep gulch among the bowlders is slipping and melting away, the trees above it are perhaps already showing a fuzzy, muffled look, the moss on the bowlders has shot its pale, pin-like points of green upward toward the sun, and the grass grows in thick tufts where the brook winds through the meadow.

CHAPTER VI

THE OPEN SEA

ONE's first impression of the open sea, gained from a steamer's deck, is usually not too happy. The mind is distracted or it is dull, even if the body be not racked, and a sorry conclusion about the sea is a common result. It is a dreary waste of waters. The horizon rim makes a perfect circle about one, the sky is a great arch overhead, and there is nothing to be seen but an occasional school of porpoises or the misty form of some sailing craft straining along the sky-line. *First impressions.*

The *nouveau* thinks the whole affair monotonous and, indeed, at first glance variety does seem lacking. Yet in reality there is not an hour when the wind does not shift the form of the waves, not an hour when the light and color of the water are not changing, not an hour from dawn to dawn when the uneasy, faceted surface is not throwing back reflections of the sky in a thousand variegated hues. The sea and the sky are always changing. What appears at first a *Sea changes.*

monotony is, in fact, an unending diversity. Time was doubtless in the infancy of the earth when the beds of the oceans were filled with pestilent gases and vapors, and time may be in the earth's old age when the seas will be great frozen depths of ice; but to-day they are in their prime, in the heyday of their glory, strong in mass and movement, overwhelming in extent and power, splendid in color and light.

Water forms.

Water at rest, like the air, would seem at first blush to be quite formless. It is the flat, even-filling of a hollow. Its positive forms are shown only when it is agitated by wind, or pushed in tides and currents, or seeking its level in lower places. There are currents in the sea, but they are hardly recognizable in the open water except by their color. Their forms are not definitely marked—not even that of the Gulf Stream—though they have certain movements, widths, and lengths, that are well known to the navigator. These currents flowing through the main body of the ocean have always called up an analogy or a likeness to human physiology. For they seem like sea arteries in their movements; and the tides rising and falling liken human lungs respiring. We are, through such resemblances, often led in a romantic way to imagine vain things about the sea; and more than once writers have pictured it as a living body — a wrinkled monster writhing in a cramped bed from which there is no escape. And the waves as they come up the rocky coast, flinging long arms upward to grapple with the rocks, have been likened to companies and legions of the deep sent to battle against the rocky barriers—companies utterly inexhaustible and gaining vantage ground always by wearing out their opponent. There is strife between land and sea, to be sure, but it is the warfare of unthinking elements and there is no enmity about it or in it. Each side is obeying the law of its nature without knowing why or wherefore. It is continuous strife, too. For the so-called legions of the sea are always marching. The "flat sea" is a misnomer. There is no such thing. At times the surface is unruffled, light and color are thrown back from it as from a burnished shield, but the shield is never motionless. Even in the tropics, where the surface may be unbroken for days at a time, there is always the great, heaving "swell" underneath. The restlessness of the sea is unceasing.

Sea strife.

Restlessness of the sea.

When the wind is rising over an unbroken

Wind and wave.

sea-surface it makes itself apparent at first in little catches or quivers on the water. The wind itself comes in fitful puffs and squalls, and it is these little inequalities of wind-pressure that make possible the breaking of the surface at the start. As the wind increases in force the surface is covered with small, facet-like waves that flash light and color with great brilliancy. With a stronger wind we have what is called a "chop sea," in which waves scurry hither and thither, driven by local gusts, crossing and breaking upon each other in small dashes of foam. If the wind is long continued from one direction the general drift of the waves and the water will be toward the opposite point of the compass.

Wave travel.

The harder and stronger the blowing of the wind, the more uniform the travel of the waves, though they are always more or less ruffled on their surfaces by eddies and contrary gusts, and occasionally a wave set in a lateral direction breaks in upon the line and churns up a great yeast of foam.

With a stiff wind the sea shows us waves crested with foam and commonly referred to as "white-caps." These caps are produced by the crest being driven faster with the wind than the body of the wave, thus losing its support; or by the crest being thrown up in the air with the upward push of the wave. The wedge-shaped cap thus dashed upward or forward breaks into spray, is filled with countless air-bubbles, and shows bluish- or greenish-white to the eye. In heavy winds this "white-cap" is apparent in every direction, but it does not break so regularly or so smoothly as in a common gale. *White-caps.*

Storm waves are usually marked by flawed and broken surfaces and their crests are ragged and torn, often being wrenched away by gusts of wind and driven across the ocean in the form of flying spray. But despite its irregularity of surface, one is never deceived about the bulk and weight of a storm wave. Its rise and heave are indicative of its power. The lift of the wave seems one long, straining effort at pushing up the gable-shaped crest. It heaves and heaves until at last, having pushed the top to an unsustainable height, it suddenly lets go as though exhausted and the crest pitches forward in foam. In long-continued storms these same waves are beaten into white, bubbling, froth-hung surfaces, foam is festooned in wreaths from every crest, and water dust rolls into every hollow; the air is full of flying spray, the clouds are obliterated, *Storm waves.*

and sky and sea seem to melt and mingle into one. But this is the hurricane storm that literally beats the sea into yeast and blurs both form and color. It is not frequently seen, and has too much chaos about it to be more than awe-inspiring by its power. It is little more enjoyable than the night-scene at sea, when rain and wind are howling through the rigging, and the white-caps gleam dull and ghost-like beside the black hulk of the vessel. Nature is sometimes too violent for either love or admiration.

The hurricane-sea.

The height of storm waves is more moderate than one would suppose. In fresh-water lakes they rise to a greater relative pitch than on the sea, because fresh water is lighter than salt water. The waves on Lake Superior, for instance, are higher in proportion to wind and water-depth than on the Mediterranean; but on neither is there any mountainous altitude attained. The heavy waves of the Mediterranean average only from thirteen to eighteen feet in the perpendicular; and on the North Atlantic, one of the most tempestuous of all seas, they are only from nineteen to forty-three feet—the latter height being the greatest ever known there. This is certainly high enough, but hardly

The height of waves.

the "mountain high" that we hear about so often from the returned tourist. It is not even hill high. It has been asserted that off the Cape of Good Hope waves one hundred and eight feet in height have been seen, but one may venture to doubt the assertion. The Cape of Good Hope region has always furnished the marvellous in sea-tales, but this one is something too wonderful for belief.

Waves "mountain high."

The breadth through or thickness of a wave is usually determined by its height considered in relation to its class or kind. On the open sea, where the friction of the sea-bottom is eliminated, the longer waves are often several hundred feet through from hollow to hollow. The long heaving swell of the tropical seas which moves under the ship, lifts it, and then passes on across the distance, its glassy surface unbroken by any dash of wave or spray, is probably the thickest of all ocean waves. The estimate has been made that it is sometimes from five to six hundred feet in its largest dimension. But this long swell belongs only to the region of the trade winds, where the push of the wind against the wave is regular and continuous. In localities of cross-winds and local storm-centres such waves

Thickness of waves.

and such thicknesses are infrequent, if, indeed, not impossible.

Wave lines. The lines of a wave made by light-and-shade and the variation of color are somewhat dependent upon wave motion. The swell of the Southern seas has about the same lines as the smaller rolls of a Dakota prairie. The horizontal ridge and its corresponding valley are distinctly marked, light-and-shade and color play all along them, and the heavens above are rolled and unrolled from them in long, flashing reflections. As soon as the surface is broken by wind the lines are blurred, and the reflection is lost in local hue, though each little wave continues to throw off from itself the tiny reflection of light and color, like a portion of a broken mirror. The general form and heave of the wave are not lost; its surface only is changed. *North Atlantic waves.* The waves on the North Atlantic are quite different from this tropical undulation. They are shorter, sharper, more ragged in surface, and they have a cross-blow tumble and toss about them that sometimes defy the line and make only flashing light and color possible. In heavy and steady winds they heap up in enormous ridges, following each other, file upon file, like other waves, but their surfaces are always irregular, owing to flaws in the wind. In fact, the only line on the North Atlantic that has any stability about it is the horizon-line—the darkest line usually on the face of the waters. Even that is not too strong, owing to the presence of vaporous atmosphere. It is only on cold, clear days that it is sharply defined.

Wave motion is more of an appearance than a reality, though there is always some movement forward in each wave, and a general drift of the water in the direction of the blowing wind. That which has real movement about it is the undulation. This movement of the undulation is very apparent in the shaking of a carpet on the lawn or the bend and roll of standing grain over which the wind moves swiftly. Neither the carpet nor the grain moves forward, but the undulation certainly does. And it often moves at a great rate of speed—say fifty miles an hour—out-stripping sometimes the winds that set it in motion, just as a heavy log in a river current when once started will move faster than the current itself. One has but to watch the movement of floating objects on the waves to be convinced that the water itself moves but

Undulation and wave motion.

slowly. A chip rises and pitches forward on a crest, but it is drawn back almost an equal distance into the succeeding hollow. Eventually it is carried many miles, to be tossed perhaps upon some island shore; but it makes a very slow passage.

Depth of the undulation.

The undulation is generally supposed to be only a surface affair—a disturbance like the ringed waves that ride shoreward from a stone cast in a pond. And so it may be; but the depth at which the movement is felt is often very great. In the bays and harbors along shore a wave four feet in height can be seen swaying and tossing the sea-weed many feet below the surface, and in the Mediterranean, where the water is very clear, the bottom of a swell has been seen rushing through rock passages twenty-five fathoms down. There is little doubt that the heaviest waves can be felt a hundred fathoms below the surface.

Local hues.

The local color of sea water is determined by its density, its depth, the ground underneath it, or foreign matter held in it. Salt water is denser and generally bluer than fresh water, and the regions of intense salinity are generally the deepest hued of all. The Mediterranean, the Red Sea, the Caribbean are at times violet-blue, while the waters near the poles, in which melted snow and ice are ingredients, appear greenish-hued. The temperature of water also has some effect upon the coloring, for certainly the warmest waters are the darkest. And, too, deep waters appear much bluer than shallow ones. The bays and harbors and coast waters generally look light-hued, possibly because of the land waters brought down and mingled with the sea, and also because of their reflecting bottoms; but chiefly because of their comparative shallowness.

The open sea on an average is about two miles deep, and in spots it is probably five or six miles; but this depth, which should, and usually does, give great body of coloring, is sometimes offset by remarkable clearness in the water. Transparency is, of course, dependent upon the mass of particles held in the water, and in this there is great inequality in the different sea areas. It is said that the bottom can be seen in the polar seas at so great a distance as seventy fathoms down. How the bottom at that depth may effect the coloring I am not able to say, but in shallow bays and harbors there is no question about the sea floor changing the coloring of the water. It is well known that certain bays with

Sea floors.

Harbor bottoms.

red mud bottoms have reddish waters and others with sandy bottoms have yellow waters. Black streaks in the water are often indicative of hidden rocks or dark masses of seaweed, and a sunken mud-bank will occasionally produce a silver-gray stripe for miles across an inlet.

Deep-sea color.

But however the bottom may change the local color in shallow waters, it has little or no effect upon the great seas. Their coloring is produced largely by particles of salt and other substances held in the water. The dust and moisture particles floating in the atmosphere are productive of the blue sky, and if we regard the waters of the sea as colored by similar phenomena, we shall not go far astray, though the analogy may not be quite exact in every way. It is doubtless the salt-particles in sea-water having the power of reflecting blue that make the Mediterranean such a dark ultramarine; and the rock-particles carried down from the Alps by the Rhone make the water of that stream assume a beautiful green-blue tone even when the reflecting blue sky is shut out by clouds. Again, the effect of the Blue Grotto, near Capri, is produced by light shining through the water from beneath and striking particles that apparently turn to blue and produce that tone throughout the cave.

Certain particles or floating matters—animal, vegetable or mineral, I know not which—make the Gulf Stream an indigo current travelling through a lighter body of water, make the Gulf of Lyons a darker blue than the sky above it, and make the Gulf of Gascony a dark green. Reference is now being made solely to local color and not to sky reflection of any kind. For if these waters be taken up in white jars the difference in hue will still be well marked. It is inherent in the water and is a part of it, just as the Yellow Sea is yellow because of vegetable deposits, and the North Sea off Scheveningen is yellow-brown from carrying in it a solution of earth matter. We can see the same local color effects in fresh-water lakes, as, for instance, in the Yellowstone region, where mineral deposits may produce red, green, blue, brown, or almost any colored water; and the warmer the water the more astonishing the coloring.

Aside from this coloring matter, the hue of ocean water is sometimes changed in spots by the presence of great swarms of animalculæ, or patches of algæ, or "sea-sawdust." The spots and areas of white, red, and brown that look so picturesque upon the surfaces of the Indian and Pacific Oceans, and occasionally in the Arctic seas, are accounted for in this way. But these are mere patches of surface-color in isolated regions. The general hue of sea-water is controlled largely by the matter of depth. It requires a great mass of air-particles to produce a blue sky, and it takes a great depth of sea-water and much reflection from salt-particles to produce " the deep blue sea." It is safe to say, then, that the greatest depths are the bluest, that the shallower depths incline to green, and the shallowest waters—the waters near shore— are the ones that show the browns, reds, or yellows.

Transparency of sea color.

All of these colors are peculiarly beautiful for a reason we seldom take into consideration— namely, their transparency. The ordinary colors of nature as shown in grass, flowers, trees, fields, mountains, are opaque. The hue is on the surface, and is only a veneer—an outer coating—so far as our eyes are concerned. But the sky in its interminable height and the sea in its vast depth are blue by virtue of superimposed layers or strata of transparent substances. It is not until stratum has been heaped upon stratum in countless numbers that the color begins to show. We see into them as into open space, the quality of the color breaks upon us slowly, and its greatest tenderness is revealed to us only in its profoundest depths.

But beautiful as the local hues of sea-water may be, they are nearly equalled by the colors that may be reflected from the surface. Light will penetrate water as it does glass, coloring it as the rays are broken and reflected by the floating particles; but like glass, water will also reflect color and light from its face with wonderful clearness. In this respect the ocean is not very different from the mountain lake and the roadside pool. The whole dark sweep of the sea brightens under the dawn and flames under the twilight, and every heaving wave is a convex mirror. Reflection is, however, conspicuously apparent only when the surface is smooth. On the glassy Southern swell it is possible to see the white clouds pass by one as in a panorama, the blue sky shaken out in great undulations, and the round, flashing sun riding the smooth waves like an enormous diamond. Whatever the sky contains will appear in the reflection. The sunsets off the Isle of Shoals in the calm evenings of August are quite as gorgeous on the water as in the heavens. Every little wave that ripples in is like liquid fire, or at times like the rounded surface of an iridescent vase. Even a

Reflections from water.

The Southern swell.

more gorgeous coloring, running at times through almost every note of the scale, is seen at twilight in the short, lapping waves off the western coast of Scotland. The Mediterranean about Spain and Algiers, the Adriatic at Venice, the seas of the Southern Pacific are wonderful in their color harmony, but in intensity of hue they seem less positive than at the North.

Dawns and sunsets far out on the open water are seldom so varied or so gorgeous as near or over the land. The air at sea is less charged with dust than moisture, and ruddiness of coloring is not perhaps so possible of attainment. Occasionally, however, in the summer months there is a sharp display of colors as the sun comes up or goes down over the water-line. I find in one of my note-books the following memoranda:

Waves of fire.

Sky effects at sea.

"July 11. Gulf Stream: The sunset colors are deep-orange, pink, and yellow, with greenish hues in the sky spaces; a fog-bank just beneath the sun is lilac-hued, turning to purple. The smooth water resplendent like a gold floor; far up the zenith the wispy cirrus clouds are shining like snow against the blue."

"July 14. Bright yellow sunset, light from the blue not very clear, yellow sun-shafts. The sun is barred with dark purple clouds. In the east, north, and south pale tints of lilac, pearl-gray, and pink."

If the ocean surface is very smooth, one may occasionally see at sunset the double sun—that is, the sun's reflection as a round, fiery light in the water, just below the sun itself on the western horizon. If the water is ruffled, we have, instead of the round light, a long flickering pathway across the waves. It takes the coloring of the sun, and is in fact only its broken reflection. When the sun is high up in the heavens and is beating down diagonally on slightly ruffled water, this pathway is less marked in color, but broader and more brilliant in light. At times when looking at it with half-closed eyes one can see, or at least imagine, the sun's rays striking the water like shot and splashing up light by the impact. The long trail of moonlight on the ocean which we all love to watch, and think about romantically as the same moonlight shining on the river at home, is a similar appearance, only the light is feebler and more mellow, and the apparent splash of the falling rays striking the water is not so noticeable.

Sunlight on the sea.

Moonlight.

Cloud-shadows are as conspicuous on ruffled water as upon the land, and the number of color-changes on the sea surface caused by clouds is little short of astonishing. Sometimes these shadows are gray, pale-green, pearly, or even brown; again, and more frequently, they are mauve-colored or lilac. The clouds producing the shadows are usually ragged forms of the cumulus drifting low in the air, though occasionally a tall tower cloud appears over a warm sea, and in periods of storm, the stratus and the nimbus. For the colors cast by these clouds upon the water I can do no better than quote my note-book again:

Cloud-shadows on water.

"July 16. English Channel: Smooth sea, blue sky dappled with white cumulus clouds, water full of color flaws. The cobalt-blues are broken by bright patches of green. The water to the east under a cloudy sky is silver-gray; the water to the west under a blue sky is intensely blue. This is sky and cloud reflection."

"July 17. Off the Solent: The water full of lilac shadows upon a pea-green sea. The clouds are low and drifting fast, the shadows shifting on the sea to correspond. A very queer color effect which one might not think possible were not the reality before him."

Colored shadows again.

The variety of colored shadow upon water is almost as great as upon land, but the repetition of similar effects is not frequent. The shadows in the Solent that July day I have never seen repeated anywhere on the water. I am disposed to think that the color in the shadow comes from the reflection of the cloud or sky, mingled with the local color of the water, but it is not possible to be certain about this.

The misty and cloudy days at sea are far from being colorless, though, of course, the sea is not so brilliant as under sunlight. Again I quote from my note-book: *Cloudy days at sea.*

"Aug. 8. Gray day, mist close in upon us like a veil, horizon-circle does not seem more than a mile in diameter. The water looks gray-green, the mass of the sea a shade darker than the mist, some green in the break of the wave. At sunset the light seen through a thin rain-sheet is very white, almost like phosphorus."

"Aug. 9. Cloudy, overcast day, sea dark, waves moderately high. The crest of the wave just below the white is a beautiful dark-green. In the churn of white along the steamer's side it is turquoise-green."

These peculiar shades of sea-green are seldom, if ever, seen under sunlight. Cloud and storm and flying scud reveal them at their best. They often appear in patches, extending over a small area of the sea, and will shift position and move off, as though caused by the shadows of flying clouds, but I have never been able to locate the clouds that produced them. Certainly they appear as the direct result of clouded light, and show at their brightest when the waves are breaking with a swash against the *The emerald greens of storm.*

black side of the steamer. They are also seen when waves are breaking on a rocky coast, and often, during storm, emerald-green is churned out of the indigo-blue of the Gulf Stream.

Northern and Southern seas.

These blues and greens, with snowy wave-crests that come with stormy seas and cloudy skies, certainly have a most stimulating beauty about them. They smack of Iceland and the aurora, and their clear, cold color suggests the crystalline purity of the sea. Quite different from such strong tones of coloring are the warm surface blues and pinks that play upon the unruffled Southern seas. The listless loveliness of light, the blend of the two vast blues, the rosy ocean of the dawn, and the golden ocean of the twilight, what a contrast to the North Atlantic! And yet, how very beautiful! From the smooth equatorial swell, all the light and warmth and glow of the heavens are reflected as in a mountain-lake. Every opaline flush upon the cloud, every pale-lilac of the horizon-vapors, every green and gold of the barred sky at sunset repeats its image in the slow-heaving wave, until the vast water seems but an inverted sky, and the whole scene in vision swims a realm of light and color.

And those soft, windless nights of the South,

the orange moon rising in the east, the smoke of the steamer trailing in dusky banner lazily behind, the black masts and yard-arms swinging slowly backward and forward across the starry heavens, the stars themselves flashing on the blue-black ocean floor! It is not possible to conjure up a more beautiful scene. The storm beauty of the Roaring Forties, yes; but ah! the great peace, the calm splendor of the Southern seas!

Following the equator

CHAPTER VII

ALONG SHORE

The restlessness of the sea shows itself nowhere more positively than where its waves encounter the opposition of the shore. The foam-backed rollers may jostle and rasp each other in the open and still drive on comparatively unscathed; but on the reef, the cliff, and the beach they fret and dash themselves to pieces. Almost every second they are breaking and falling, but their number seems not to lessen. New ranks replace the broken vanguard; the breakers are never quiet, never at rest. On sand and beach and promontory the rub of the water is always felt, the wash of the wave is always heard.

In calm weather the gentle, smooth-tongued swells seem quite harmless as they play in and out of rock-fissures and gravel-pens, or fall lightly on the white sand of the beach; but it is quite a different tale when the storm waves break, booming and crashing, on the coast. The battering power of water is enormous, and it always works more destruction on a shallow coast than on a deep one. A cliff, for instance, that has a shelving bottom leading up to it has much more fury directed against it than one with a base seated in deep water. The waves will not rush forward and dash into spray against the latter. On the contrary, they flood up heavily and slowly, and seem to stop without striking a blow, the crests dancing up against each other rather than against the rocks. The reasons for this will be apparent if we consider for a moment the conditions that make possible the "breaker" and the "beach-comber." *Along the cliff.*

The surf breaks most violently over shoals or along a shelving beach where the bottom of the ocean bevels upward toward the shore. A wave from the sea is pushed up this incline with a swiftness proportionate to the propelling force behind it. The friction or drag upon the wave comes from the shelving bottom; and as the shore is neared this friction is not only intensified by the increased abruptness of the incline, but also by the flow outward of those returning waters from the beach which we call the "undertow." The result is, the *Why waves break.*

lower part of the wave is not able to keep up with the upper part, the top is shot violently forward, and having no base to rest upon, breaks and falls upon the beach in spray. The cause of the dancing upward of the waves under the deep-based cliff will now be apparent. The base of the wave, meeting with no marked friction, moves as fast as the top and strikes the rock beneath the surface. The whole wave rebounds against the wave following it, and a push upward of the water in vertical points or dancing jets is the result. There is no other way for the water to move.

Dancing jets.

It will not have escaped the notice of the most casual observer that the waves breaking upon a coast follow each other more closely than upon the open sea. The friction upon the waves as they reach shallow water—the drag upon the bottom—is also responsible for this. The front ones cannot move so fast as the rear ones, and there is a closing up of the ranks—sometimes a doubling or tripling of the waves. This at times results in the waves along shore being smaller than on the open sea, and again, in times of storm, it may result in their being larger. Certainly a storm on a rocky coast will throw the breaker-crest higher than upon open

The size of coast waves.

water. For a different set of forces regulates the form and motion of the crest in mid-ocean. The white-cap on the open sea is lifted to a height where it cannot sustain itself by the push up of the water and the wind ; but the wave has no beach beneath it to concentrate strength in the cap. Driving upon the coast, the cap is flung forward by the wedging process already described, and, if there is a fierce storm, it is often shot up the shore to a great height. Light-houses on rocky ledges far above the sea-level have been frequently washed over and destroyed by these enormous breakers, and upon the cliffs of the Irish coast the waves sometimes rush up fully two hundred feet. The blow struck upon the cliffs by such masses of water is estimated at from two to three tons to the square foot; and a mile back from the shore the ground can be felt to tremble under the terrific impact. It is the sharp, upward incline of the shore bottom that makes such waves possible. On the open sea they could not by any chance rise to such a height. The maximum of the Atlantic wave has already been given at forty-three feet, and not even in the Roaring Forties, in the most violent storm ever known to roaring sea-captains, has a

The power of the waves.

higher wave been known. It is only by forcing water on a coast or through a channel that great height is attained. The fifty feet of tide that rises so rapidly in the narrow, wedge-shaped Bay of Fundy is an analogous illustration.

Forced waves.

The beach-comber, as it comes in upon the sloping sand, is somewhat tame compared with the rocky coast-breaker. It rises and falls with more apparent regularity. As it rises, a long line of light may be seen shining beneath the top, and the already curling crest seems hurrying down to meet it. Just at this time the wave shows its greatest beauty of color. The foam of the top is half water, half air, and is bluish-white, while the green and dark-blue of the wave are the more transparent for the light shot through the thin concave of water. The glassy curve shows for a moment a whirling panorama of beach, sun, and sky; the base of the wave swings back and under, the crest swings over, and in another moment the whole structure has broken in a froth of foam on the shore.

The beach-comber.

If the beach is sandy and quite flat the broken wave pushes its waters in a gentle flood upward and outward in rings and half-circles, edged with white-beaded foam; and these, as they advance and pause for a few seconds, take on wonderful forms and still more wonderful colors. Especially brilliant are these flat mirrors at evening, when the waves are not running high and the heavens are bright with sunset hues. The reflection is more delicate than the sky overhead, and the colors melted and fused on the glassy surface, run together with a harmony beyond analysis. Every hue and tint are there, and all are softened and warmed by being seen in the watery mirror.

Water-mirrors on the beach.

But the water pushed up on the beach lingers for only the fraction of a minute, and then slowly turns and runs back under the base of the newly forming wave. Some of it runs out into deep water in the undertow, but the bulk of it helps form the base of another wave. It will be remembered that waves themselves travel but slowly, and that the undulation furnishes most of the movement. It is not new water that comes in with each wave. If it were, one might wonder what became of the old water fallen upon the beach. The little current there is in the undertow would not be sufficient to carry it off, and besides, the running of the undertow is not always apparent. It is the

The undulation again.

undulation that makes the breaker, and the constant swash of waters on the beach is as the fringe of a cloth flapping in the wind. When great quantities of water (not undulation) are driven in upon the shore by heavy and continued wind, the sea rises and floods all the inlets and marshes; but in the falling of waves upon the beach there is no rising of the sea. The beach-combers are made up of substantially the same water cast into new forms, new lights, new colors.

The thrust of waves.

The lateral direction of a reef or beach has little to do with the direction the waves may take. It may retard or cripple their force, but it has slight influence in turning them aside or making them follow another course. There is not enough cohesive body about water to have its course turned except by slow degrees. That which gives the "set onward" of the waves is the prevailing wind, and, once started in a certain direction, the waves run on until broken to pieces against the rocks or the beach. And it is interesting, perhaps, to know that the waves seldom strike the coast or the beach a full broadside. Instead of coming straight on they are usually a little twisted, so that they strike the beach at an angle, and the travel of the breaking crest may often be followed by the eye for a long distance down the shore. This side thrust of the waves has one very positive effect. It wears away the shore, and that, too, faster than any direct blow. The side push works in swirls, sand is swept along the beach and gradually dragged into the sea by the under-currents to be carried off and deposited on some near-by shoal or bar, and the tendency is to hollow out the beach-line in half-circles. As a result we have the beautiful sickle-moon curves that mark the sand-beaches on almost every sea-coast. Next to the lines of the snow-drifts, they are as graceful, perhaps, as anything the eye may see—save always the lines of a flowing river. The best place to see them is from a high cliff, looking down along the shore. The curves of bay and beach will then appear quite perfect. *Wear upon the beach.* *Curves of sand-beaches.*

The same form of wave-action works similar results upon the rocks of a coast, but with less ease and uniformity. The water, striking full-faced against the rock surfaces, is simply shattered into foam, but coming diagonally it gains cutting power by a rasp and a grind all along the bases. And in this grind the loose stones and bowlders, hurled and rolled *Wave-action on the rocks.*

along, one over another, prove very effective weapons of destruction. They are swept in and out of pools and crevices, lodged in pot-holes, caverns, and scoops, and churned round and round by eddies and currents with a scrape and a grate at every turn. The cliff is thus gradually undermined, and needs only the heave of frost to topple it into the sea. It is protected in a way by its own ruins—the outer guard of fallen rock that breaks the force of the waves—and besides this, the cliff bases, as well as the fallen bowlders, are sheathed with fringes of seaweed and barnacles; but still this grind of the surge and the ceaseless beat of the surf finally wear all of them away, and their particles, like the sands from the beaches, are carried out to sea by the under-currents and deposited on the shoals. The diagonal thrust of the waves has something of the effect upon the shore that the running stream has upon its banks. It not only has cutting and wearing power, but it makes currents which carry off what is cut away.

The greatest wear of the waves is, naturally, where the rock is the softest. A hard quality of rock—so hard that it has endured—usually appears as the armored prow of every project-

[margin: Cliff undermining.]

ing cape or V-shaped promontory that stretches out into the sea. It is the outlying guard, and so long as it stands it protects what is behind it. When the sea finally wears away the point it is likely to leave a sunken base but a few feet below the surface, over which the waves break in spray; or perhaps there remains one of those fantastic pinnacles or pillars, usually called Devil's Pulpits, which may be seen along almost any rocky coast. At times again, waves wearing upon a soft portion of a rock hollow out caverns or perhaps passages clear through the promontory, into which the water rushes and issues on the other side in a tumult of spray. When the supporting sides of the cavern are of sturdy material, the roof may remain after the rest of the promontory has been eaten through, in which case we have the natural bridge or arch—a not infrequent sight on rocky coasts, and certainly a picturesque one. A more common way, however, of wearing the rock is by the water following the seams and cleavages opened by frost. The savage thrust of the sea through these cracks sometimes results in the "spouting-horn," which flings up its jet of foam with great force, and under sunlight with surprising beauty of effect. Still more common is the

Rock forms made by water.

Pulpits, bridges, and caverns.

grind of wave and bowlder on the base of the cliff, until it is so eaten away that the top heaved by frost falls into the sea by its own weight. In either or any case, and however the wear may take place, it is slow annihilation for the cliff. The sea gains inch by inch.

The forming of sand-dunes. But the shore is not subject to all loss and no gain. Occasionally a great storm brings sand in and heaps it up along the beach. This is the beginning of the sand-dune—the great protector of the land against the sea. It must not be conjectured, however, that the high dunes of the Cape Cod shore or the low sand-banks of the New Jersey coast are wholly the heaped-up deposits of the waves. Dry sand will drift with the wind very much like hard ball snow, as anyone who has been on Sahara will testify. Even the tourists at Cairo, who never go beyond the Mooski, will be able to say how many times the Sphinx has been dug out of the drifted sands of Egypt. Along the exposed shore, where the winds are always restless, the loose sands are kept in continual motion, and it is the winds that round up and build the hills and valleys of the sand-dunes. In addition to the sands brought in by the sea, the land breezes drift quantities of them down to the shore when the tide is out, and pile them over the tops of broken masts, sea-weed, and rubbish, until they make a bank that may be a barrier against the wave. Once a mound is made it is held in place by the hardy grasses and scrub-vegetation that grow on its sides and top. The mound or bank may keep adding to itself in this way until it stands a hundred feet high along the coast, and makes an almost impregnable sea-wall. Behind such protection as these sand-dunes, and by the building of dikes across the ocean inlets, thousands of acres of water have been turned into green pasture-lands and flat fields. Holland is an illustration of it. *Sea-barriers.*

There is still another way by which the land gains upon the sea. The rivers, coming from a great distance inland, flood drift and dirt into the ocean. After a time the mouth of the river begins to choke up with muddy deposits; a bar and then a bayou or lagoon is formed, a marsh begins to rise above the waters, seaweed accumulates, rushes and flags spring into life and make the ground stable; and after many years a group of islands, or perhaps a habitable meadow-land, is formed. Venice was builded upon such a formation, caused by *Bars, lagoons, and marshes.*

the deposits of the Brenta; and the original bar that caused the choking of the river and made the lagoons is now called the Lido.

<small>*The tides.*</small> The tides in their rise and fall have some effect upon the shores, and they wear more or less upon the harbors and narrow inlets; but their usual comings and goings are too pacific to cause much injury. On the coast of Florida, where the tide rises only about a foot, there is no appreciable effect, but it is quite different in the Bay of Fundy, where it rises sometimes fifty feet, and with great rapidity. Its wear in that arrow-pointed bay is almost as severe as that of storm waves. Along the shore the rocks are worn horizontally, and show in jagged ledges like strata of slate.

<small>*The flood-tide.*</small> But the Bay of Fundy is rather an exceptional case. Usually the flooding of the tide is a noiseless stealing upward and inward of great bodies of water. It backs up the river, rises through the marshes and meadows, covers the reefs, bars, and beaches, and hides from view the sea-weeds, the barnacled rocks, and the shattered hulks of sand-sunk ships along the shore. It is well called a "flood-tide," for it is little more than an inundation of the sea. It is interesting to watch as it creeps and

spreads, and it may stir romantic thoughts in the minds of lovers; but from a picturesque point of view one is at some loss to discover anything remarkable about its looks.

Not so with the ebb-tide, when the water goes out and leaves great beds of rock and sand and reef exposed to view. It is not merely that the exposed places are curious for their wealth of sea-weed and barnacle and stranded ocean-life, but they are often extremely interesting as form and color. The great bowlders covered with clinging fringes of sea-weed are graceful in outline, and quite charming in such tones as dull yellow and sage-green. The pools left in the rocks and the gravel-pens are marvellous studies in different hues, and the dark, water-worn rock-bases offer a strong contrast to the light-gray tops bathed in the sunlight. Even the black spots of sunken ledges, hulks, or broken piers that peep above the water at low tide have a picturesque quality about them, lending accent to the scene; and the sweeping indentations of the coast, the bridges, pulpits, and rugged promontories all seem so much more powerful and massive in form when the tide is out.

The ebb tide.

The curves and lines along a coast are an unending study. Not merely the smooth bend of a basin worn out by water, but the vast, rounded curve of a headland against the sky; not merely the graceful roll of wave-worn lines on a sand-beach, but the circular swing of a cove or bay against the sea. Lest there should be too much flowing smoothness about such lines, there is always the rectangular block or the splintered shaft that protrudes above the line of the headland, the uneven sand-dune, or the broken mass of the forest running back of the bay to act as a foil. These are the sharply accented marks that save the scene from weakness. And the broad masses of color are not less powerful. The cobalt-blue of the sea turning to violet in twilight shadow, the white, gray, or yellow of the shore, the deep greens of the forest, the blues and whites of the firmament—where else can such colors be equalled? They are the primary chords in one of nature's greatest harmonies.

The coloring of the coast is more susceptible to the influence of light and sky reflection than almost any other portion of the earth. Possibly the cause for this lies in the great reflecting field of water so close at hand. The sea not only throws back the light of the heavens, but it thickens the coast atmosphere, thus regulat-

Coast lines.

Masses of color.

ing the distribution of light, and tingeing the whole coloring of the shore. And the shore is so very easily tinged. The pebbles, shells, and mica sands that go to make the beach, whether wet or dry, respond in coloring to the light. Even the rocks are mellowed by it. Close at hand they may look yellow, brown, or gray, according to their mineral composition. Along the New England coast they are dull yellow, stained with iron-rust, and if one of the little pools lying in a hollow of a rock be examined it will disclose a background of bright orange; but this local color is not apparent when a jutting headland is seen from a distance. A gray light may blend into sobriety the colors of the cliff, the white beach, the dark pines, or a warm-yellow sunlight may enliven them all with a new hue. At twilight pink and rose may spread from sky to water, and from water to sand and rock, until the whole vision is a rosy one. At other times the scene may show a golden, a greenish, or a bluish tinge, dependent always on the light.

Light upon the shore.

Twilight colorings.

Beautiful by day the shore is perhaps even more beautiful, certainly more impressive, by night. The moonlight silvers the tall cliffs until they look like vast fortresses of marble, and the sand of the beach gleams white as winter snow. The Fairies' Pathway of moonbeams, or as the Chinese call it, the Golden Dragon, twists and flashes upon the eastern water, the dark pines stand in silent ranks their tops spread against the purple western sky, and from the dividing line of land and sea comes that eternal surge of the wave. How it hushes the cry of the mortal—that sullen moan of waters! What human woe or weariness but sobs itself to sleep at last! But for the sea there is no rest. Under the stars, as under the sun, to-day as through the long centuries of yesterday, it throbs and beats at the feet of the earth, and its voice is never stilled.

And is not the sea-shore equally beautiful in storm, when the spray is flying high above the cliffs and the rock-bases are trembling with the shock of water? The majority of us see the coast in the calm months of summer when it is not agitated by long storms, when the life-saving service men have closed their stations, and only the curl of the breeze-wave is seen on the beach. But the time when the sea is in its full power is mid-winter, when the land is white with snow and the wave is white with foam. Then the roar and hurly-burly of the waters beating against the cliffs and crashing on the beach, tell us what latent strength lies in our whilom summer sea. The pound of the waves is terrific. They rush and dash along the ledges and through the fissures, and are flung high in air by every stubborn headland. After many hours of this wild charging the water itself begins to have a beaten and battered look about it. It is churned about the rocks until it hangs in ropes or skeins of foam, and oftentimes the very whiteness is whipped out of it—the froth lying in soiled, cream-colored streaks upon the surface. In such storms many a heavy block of granite is shaken from the cliff-wall and rolled into the sea, and many a new inlet or bay is cut out in a few hours by the steady beat and wash of ponderous breakers. As for the fate of a ship driven on a bar or shoal in such a storm, it can readily be imagined. As soon as she strikes the sands the waves begin to break over her decks, and a few hours may suffice to see her stove in and pounded to pieces.

The whipped waves.

On the bar

It would seem as though this destruction of cliff and beach were anything but a blessing, and yet the storm at sea has its uses. If it harries and worries the shore, it helps the broad acres lying back from it. Out of the ocean come the vapors that form the clouds; and the massive ranks of nimbus that voyage inland with the storm, creating uproar all along the coast, are the water-carriers for the land. The fountain, the stream, the brook, the river, and the lake; the dew on the grass, the sap in the tree, the color of the flower, and all the gorgeous garmenting of creation, are due to the vapors of the sea. If the time ever comes when there shall be "no more sea," then will come with it an end of all life. The primary physical conditions of life here on earth are heat, light, and moisture. With the last element gone, the first would follow, and the second would be rendered useless. The world would be as cold, dead, and colorless as our skeleton satellite the moon. The dread sea—so-called—was not created in vain. It has its uses and it certainly has its beauties. *Mare horrendum* it may be to some; but to those who know it well and have lived upon it or beside it all their lives, it is as lovable in its stern character and majestic desolation as the sands of Sahara to the wandering Bedouin, or the tumbled-and-tossed Bad Lands of Dakota to the predatory Sioux.

CHAPTER VIII

RUNNING WATERS

It is seldom that a river empties itself into the sea from between high banks of earth or rock. Long before tide-water is reached, the banks have usually fallen back and away from the stream, the course is through undulating country, flat plain, meadow, or marsh, and the stream itself in the last few miles of its run usually flattens out and becomes shallow. About the mouth or mouths, for there are often several of them, are heavy deposits of mud and sand which year by year the stream has been carrying down; and these choke and raise the exit, causing the water to move slower. As it nears sea-level its velocity and its wash are perceptibly lessened, its course is tortuous like that of a wounded snake, and its very slowness is favorable to the settling of its sedimental mud and sand. At last, when the stream reaches the sea, its final leap of mad freedom into its ocean bed is less apparent in the reality than in the imag-

The river at the sea.

ination of some graphic narrator. All the rivers I have known melt into the sea as smoke fades into the air. The current is loosed from its confining banks, but it still holds headway out upon the top of the salt water for some distance, its coloring marking its course, until gradually it breaks into thin, cloud-like sheets and is finally absorbed and neutralized by the vast body of the sea.

The sluggish flow.

If we enter a river from the sea, we may have some difficulty at the start in finding the main stream. The water is spread wide, and there are many false inlets and bayous scattered here and there. Even when we are at last in the main channel, we find the water discolored and moving sluggishly between low, ill-defined banks. There is little movement at this final stage of river life, little winding in and out of nooks and bends. The stream seems to drift and drag lazily along, with none of its mountain brightness. It is moving slowly toward annihilation, and it seems almost semi-human in a consciousness of it. Farther inland it flows a little freer and has more power. The

Through the meadows.

salt meadows stretch out on either side of it, and the banks have lifted, perhaps, several feet in height. These banks are formed of mud and held firmly by the roots of flags and grasses, their edges are ragged and under-cut by water, and upon them occasionally grow small elms or clumps of briars and alders. The bed of the river is muddy, the water cloudy, the color of it beautiful only in sky reflection.

As we ascend the river, following what is called its Plain Track, the banks continue to rise, the bed becomes sandy or pebble-strewn, the stream clearer, the character of the ground more substantial. There are lifts or rises in the land, that seem to indicate little hills that have been worn down by many centuries of water-wear; and these are, in fact, the forerunners of the hills which we soon find rising on either side of the river. A hundred miles or more up the stream, the hills begin to jut out stronger. They may be near at hand, but more often they are several miles back from the banks, and the river-bed is a flat plain lying in between them. The land may be cut up into farms, with fields of grain, orchards, and white houses; there may be forests here and there that grow down to the water's edge, and meadows where cattle roam and daisies grow, with fords, bridges, and occasionally a lonely mill. The water does not run swiftly as yet, but it winds and cuts in the

The Plain Track.

The river basin.

muddy banks and goes a long way out of its course to get around a piece of hard ground. It is deep in places, too, and has a lazy fashion of sleeping in flat pools under the shade of some great oak or elm. It is in no hurry to be gone, and yet it always keeps moving, drifting seaward.

The Valley Track.

We meet with quite a change in river character when what is called the Valley Track is reached. It seems as though the great plain had been narrowed, as though the distant hills had grown almost to mountains and stood closer to the water's edge, and the flat farms had been converted into side-hills or foot-hills. How different now is the river! It has a rocky or stony bed, there are sharp, confining banks; sometimes there are cliffs, about the bases of which the clear water laps and gurgles. The stream is now running swiftly and turns in bends and angles, flashing light and color from its rippling surface. There are also rapids at different places, and where the bank bends sharply we meet with racing water on one side, and the deep pool with its back-water on the other side. Clumps of saplings or dank masses of bushes fringe the sides and droop into the water, and occasionally in the centre of the stream is a long, thin island of earth and rocks, its top capped with pines, and its shores fringed with willows turning their silvery leaves in the wind. The prow of the island, so to speak, is usually of hard rock or compact gravel, and it seems to cleave the river in twain, leaving the two halves to spin away on either side, much as the waters seem to hurry by the sides of a great ship at full speed.

The river island.

And how the river does sweep along this Valley Track! It does not babble and chatter, or pitch and toss, like a shallow brook, yet it is merely the brook come to maturity and sobered by mass and volume. Its murmur is hoarser, its bed smoother, its course less interrupted; yet still the life of it is in its movement. Sweep and glide, sweep and glide! In and out of bend and basin, around and about rocks and islands, now fast, now slow, now complaining over shallows, now soundless over depths, regardless of obstacles or difficulties, it keeps moving, keeps moving. In storm and calm, under sun, moon, and stars, the flow is forever slipping seaward.

Hurrying water.

One would hardly suspect that the smooth, lapping waves that feel so soft to the hand trailed in the water from the side of a canoe—those waves that glitter so innocently in the

The wear of water.

sunlight—have a cutting and a wearing power that nothing can withstand. There is no edge to water itself, but its action sets grit and gravel, stones and even bowlders moving, and the teeth of these are very sharp. A stream running four miles an hour will roll down stones nearly three inches in diameter, and wherever the water flows and particles touch, there is wear upon the land. This never-ceasing rub, rub, rub, carves deep lines in the course of centuries; and so it is that the smooth water becomes the great sculptor of the earth. Standing on Storm King and viewing the valley of the Hudson, standing on the Minnesota bluffs and overlooking the valley of the Mississippi, standing on the heights above the Cañon of the Colorado, we gain some idea of what lines

The sculptor of the land.

this great sculptor can cut. A gulch five hundred or a thousand feet deep, from one to ten miles broad, and from a thousand to two thousand miles long, is not an extraordinary feat for water to accomplish. Along the sand-stone battlements of the Mississippi bluffs, far above the present bed of the river, the trace of water-wear is still plainly visible; and centuries ago the little hills, the inland valleys, the clefts and cloves and narrow defiles in the Catskill Mountains, were hollowed out and rounded by the constant touch of running streams. In all countries and along all rivers the waters have smoothed and rubbed and polished the sharp points and jutting promontories; through many centuries they have cut and worn away and modelled anew the mountains and the valleys, until to-day we have as a result those sweeping lines of beauty which mark not the Hudson alone, but the Seine, the Rhine, and the Danube.

Valley carvings.

These great carvings of the earth's surface were probably never witnessed by any one generation, or even race, of men. The work was wrought gradually, and yet, within the river's bed, one can see evidences of erosion going on to-day. Water has not lost its cutting power. If it always ran straight it would work less destruction; but the river is very susceptible to influences, and swings first to one side and then to another side, much like the pendulum of a clock. A current shunted over against one bank rebounds upon the opposite bank lower down; and a violent push given to the water by a rocky cliff may often be felt in oscillations for miles down the stream. It is this bound and rebound, from shore to shore, with its con-

Oscillations of the stream.

sequent friction, that washes away the banks, and though nature has a way of protecting the loose earth by growing vegetation upon it close down to the water's edge, yet this does not entirely save it from wear, nor the bed of the river from shifting place.

Lines of bank and stream.

It is the cutting away of the banks, the making of crescent curves and long serpentine bends that give the river some of its most picturesque features. The lines of the shores are but repetitions of the water itself, and for every high cliff that breaks the flow of the shore there is a dash and a turmoil of water that break the downward sweep of the stream. These river-lines are never seen so distinctly under the foliage of summer as under the snow of winter. The snow muffles and covers everything to the water's edge. Hill and valley, bush and tree, bowlder and beach, the overhang of the bank, the abruptness of the river island, are all smoothed into graceful contours. Upon this white background the dark-looking water dances and flashes, swirls and ripples; and the unbroken harmony of the lines, the continuity of the movement are things of beauty unsurpassed by nature in any of her creations.

The summer foliage blurs the graceful cutting of the banks, but compensates for this loss by a wealth of color. The stream sparkles between great borders of green, reflecting the blue sky where smooth, and turning to amethyst where it runs over shallows. The tree and the bank, the fern and the burning cardinal flower are mirrored in the dark pools, the cloud shadow and the sunburst are flung across the moving surface, and the path of the moonlight weaves and ravels there as on the sea. Flexible and changeable as the sky above it, the river glides along, and, chameleon-like, takes its color from its surroundings. It may be whipped with rain-squalls to-night, but to-morrow it will show the first silvery light of dawn upon its shining face, and whatever momentary effect may mar its surface, there is no pause in the smooth slipping seaward. *Color on the river.*

Even in winter, when the river is covered with ice, the murmur of the water beneath says it is still moving toward the ocean. Its face is masked, its color is gone, even its reflection is dimmed, for ice unless very smooth is a poor reflector; yet still for all its desolate state and the cold, dark ranks of trees standing along its banks, the beauty of the river has not entirely departed. *Under ice.*

The green hue of ice, the blue of the snow shadows, the glisten of the white particles in full light, are brilliant; and when the ice breaks and goes jostling down the stream it very often piles up into fantastic masses that are beautiful in color when struck by the sun.

Freshets.

A great change comes over the stream when the murmur of its water becomes the surge of an inundating freshet, but it is not a change for the better. The river itself is lost in the flood, and both its channel and its character are temporarily obliterated. A freshet, such as frequently covers the Mississippi "bottoms" from bluff to bluff, is interesting perhaps, but hardly beautiful to look upon. It is only a mad rush of muddy water. All the streams of the watershed are swollen beyond their banks, and pour into the river a turbid mass of water filled with all sorts of earth, driftwood, uprooted trees, and the like. The sweep downward of the flood, the danger, the destruction, may prove attractive to some, but the general impression upon the average person is rather dreary. A freshet in the Missouri and the Yellowstone is still more dreary and dirty, since it is nothing but a solution of mud which soils everything with which it comes in contact.

Floods.

The cloudiness of the Missouri is the natural result of its draining the alkaline plains; but the turbid condition of many large rivers can be traced directly to civilization, the axe, and the plough.

In its normal condition, and as it appeared thirty years ago, the sun never shone on a more beautiful river than the Upper Mississippi. Then the tall bluffs along the stream were covered with timber, the bottom-lands were a mass of tropical undergrowth out of which rose majestic elms, oaks, maples, and sycamores; the river itself was clear and wound its bright way over sand-bars and by many little islands. There were no railways stretching along the shores, and the small towns that stood by the river's banks had hardly made an impression upon the wilderness. All was quite as wild and primeval as one could wish, and every traveller standing on the deck of the river steamer, as he ascended that stream felt the freshness of the air, the brightness of the light, the unmarred, the unbroken beauty of forest, bluff, and shining water. A beautiful river it was, and never more impressive than at night in storm when the pilot at the wheel was finding the channel-way by lightning flashes, and

The Upper Mississippi

The river to-day.

the great elms in the bottoms and the oaks on the bluffs were roaring with the rush of winds. There is still some charm of wildness left about it but its primitive glory has departed. The tall timber is gone, the back-lying prairies have known the plough, the tributary streams draining the broken ground run mud, and there is little purity now in the water that flows to the Mexican Gulf. Years ago the division line between the clear waters of the Mississippi and the clouded waters of Missouri, where they met at St. Louis, could be traced for miles, but now one stream is about as turbid as the other. Man is the prince of destroyers, and if there is one spot above all others where he has fairly revelled in destruction it is western North America.

But all the destruction and all the muddy rivers are not ours. The Hudson, the Susquehanna, the Connecticut, and many other American rivers are still comparatively pure. And there are fouled rivers in other countries. I have vivid memories of different summers spent beside the Thames, the Seine, the Rhine, the Danube, and the Arno. The Danube and the Rhine are always referred to as "blue" by the poets and the guidebook makers, but I never saw either of them that hue. They are usually a drab color, and sometimes after rain, yellowish or brownish. In local hue they are not attractive to look upon, but muddy water does not make a bad reflector of the sky. Indeed, the Rhine and the Seine are often beautiful in their reflections and show us many odd, amalgamated colors. For clouded water will not reflect the same hues as clear water. Even the brown-hued water in the wood-lakes of America will darken the green of the overhanging leaves in reflection, and make the white flower of the dogwood appear of a grayish tone ; and a muddy, yellow-hued river like the Tiber will sometimes cast pinkish reflections and occasionally toss up little crests that appear cream-white. The Thames, too, takes on an infinite variety of colors under different lights; and in cloudy weather the Arno is just as fitful, just as changeable.

European rivers.

River reflection.

There is a third stage in the river's course remaining to be traced—the Mountain Track. It is usually called the first stage, but for the sake of convenience we are following up from the sea and reversing the order. In its Valley and Plain Tracks the river remains a river, but in its mountain course it is usually little more than a brook. And a brook is, or at least may be, a river in miniature. It usually comes from the hills, but it may come from an upland lake and creep across a flat meadow in a stupid way, lying lazily under bridges and making pools for cattle and ducks at every bend. Again, it may wind down through some heavily timbered country, its passage impeded by driftwood and fallen logs (like so many of the Adirondack streams), with little beauty to commend it save its golden-brown coloring taken from decayed vegetation. Still again, it may come off the moors and flow through the peatbeds of Scotland on its way to some loch, passing by great bowlders in the bed and scrub-timber on the banks, without being strikingly beautiful save in the ale-like hue of the water after a heavy rain.

But none of these brooks quite realizes our idea of a mountain-stream. The true brook is to be found in the Catskills, in the Berkshires, sometimes in the Alleghanies, the Blue Ridge, or the Rocky Mountains. The local community usually gives it the commonplace but descriptive name of "Clear Brook" or "Stony Brook." At its mouth it often joins the river, much as the river joins the sea—that is, with

some heaviness of movement—but higher up
in the hills it is all rush, vivacity, and sparkle.
It chatters and gurgles and swishes and swirls
all day long, working its way in and out, over
and under bed bowlders, waterfalls, and deep
pools. Where it runs through meadow or low-
land, it keeps changing and moving its banks
continually. Like the larger stream, the swing
in of the water toward the shore hollows out a
pool or deep eddy, and the sand removed from
that bank is always deposited a few yards below
and on the opposite bank, where a bar is forming.

*The moun-
tain-
stream.*

This shift of bed is not so noticeable farther
up in the hills, where the brook runs between
shores of rock. The change in the confining
banks is slight, but now there is wear of another sort. The waterfall keeps cutting back
into the rock, the pool or basin beneath the
fall keeps deepening, the bed along which sand
and stones are hurried keeps sinking, and the
vegetation year by year creeps lower down to
cover the bare shores left by the receding water.
The erosion of the brook tends toward deepening the ravine and producing what is called
the gorge or the glen. The wear here is, in
proportion, the wear of the whole basin from

*In the
ravine.*

mountain to ocean. The water cuts its bed with rolling rocks, and the rocks themselves that fall into the brook as bowlders are ground to sand and silt before they reach the sea. The stream is a grinding mill in every part of it, and no wonder that the bed cuts back and cuts deep in the glen where the current runs so swiftly.

The gorge.

The mountain-brook with its dash and flash, its abrupt banks, its overhanging foliage so cool and quiet in the heat of summer, is the most delightful of all nature studies. Especially do we find it so if we come upon it fresh from months of living in the city and spend our first day of vacation tracing the water to its source. Every feature of it seems so fresh, so instinct with life. The stream in its irregular bed twisting about among bowlders, the rocky dripping banks covered with mosses, twining vines, and rank ferns, the break of sunlight through the foliage, how very beautiful they all seem! On such a day, in such a place, the joy of being alive—of simply breathing, seeing, hearing, touching—is intense. How long we stand looking at the shiver and tremble of the water running over a flat rock! How long we sit beside the waterfall watching the plunge of the brook

Following the brook.

into the dark pool where the trout lie! The reflection of the trees, the delicacy of the transparent sky, the light, the shade, the flashing line of the brook far down the glen, what do they not say to us of life and beauty! Very pretty in its bend, very lovely in its light and color, is the water of the fall as it is pushed out and over its ledge of rock into the air. If it has no great pitch down, its curve is unbroken. Where it begins to bend there is a bar of sunlight running across it bright as silver, which changes only with the sun, and where it plunges into the pool there is a dizzy dance of bubbles coming and going as tiny spots of light.

By the waterfall.

This little waterfall, so delicate in its play, we may watch for hours, and afterward hear its low murmur in our ears whenever we choose to think about it; but its charm soon vanishes when it becomes a cataract. Sometimes the descent of the fall is so great, as in the case of many Yellowstone and Yosemite streams, that the water is blown out and shattered into mist before it reaches the ground. That seems to be in a way mere annihilation. The Staubbach, in the valley of Lauterbrunnen, is thus practically destroyed. Its wave through the air in falling is graceful and is much admired; but I am

The cataract.

frank to confess that it always impressed me as one of nature's lamentable accidents. I am also frank to confess that no great waterfall or cataract ever gave me anything but a cold chill. Niagara is merely a great horror of nature like a lava-stream pouring into the sea, or a volcanic explosion like that of Krakatoa. Grand it is in its mass, and sometimes beautiful in the coloring of the rising spray shot with sunlight; but its chief impression is one of power unrestrained and catastrophe unavoidable. It is nothing less than nature committing suicide.

<small>*Niagara.*</small>

The Catskill or the New England brook is perhaps the most enjoyable of all the small streams, because of its purity, its wildness, its tangled undergrowth, and its vivacious motion. It has many beauties of line and also countless varieties of color. Not the greens of tree and grass and moss, not the glow of mountain-flowers or the flare of autumn foliage, not the blue-and-white of sky-patches—not any of these alone; but all of them together, mingled in the delicate reflections of the brook water. The local color of the stream and the color of the objects reflected struggle for mastery. Sometimes one conquers and sometimes the other; but more often they make a surface-compromise, each

<small>*Brook reflections.*</small>

giving up something to form a compound that is the coloring neither of the sky nor of the brook, but a beautiful blend of both.

In the winter the brook is ice-bound, and its only sound is the gurgle that tells where the water is still running away to the sea. The first fall of snow in the glen, when the hemlock branches hang heavy, and the fern and bowlders lie white and still beside the dark running brook, certainly produces the picturesque; but after the water freezes and the snows deepen, the charm of the brook has flown. It is seen at its best in the hot months of summer, when the moss is thick on the rocks and the shadows are dark on the pool.

The frozen brook.

Purity is always the essence of the small stream, but purity is an impossibility where the drained surface is not rocky or sandy. The rapid run of water over clay or black loam can produce only muddiness. Such is the result in the brooks that come down from the alkaline plains of the West, and in many of the streams emptying into the Mississippi and the Ohio. The brooks of Scotland draining the peat-beds and heather are naturally dark, but running in rocky channels they have a tendency to clear themselves. The Swiss and German brooks are generally bright and clear, and not unlike the Catskill stream. Indeed, I am not so patriotic that I would arrogate all purity to my own country. I have described the Catskill brook only because it is typical of the river on its Mountain Track. Fortunately there are many streams like it on the face of the globe.

Purity of brook water.

The source of a stream is often the cause of some disappointment to the finder of it. Sometimes it fulfils expectation, and is a small basin of bubbling water rising from beneath a huge rock. Its overflow forms the rivulet that finally develops into the brook. The water in such a case usually comes from a subterranean spring and flows cold and clear, following some vein or fissure in the rock. In Scotland the source is usually a "well-eye," as in Switzerland a glacier; but in America the beginning of the stream is not always so simple or so poetic. Many of the brooks when traced to their origin are found to come from small lakes fed by subterranean springs, or more often from a weedy, rush-grown marsh, which acts as a catch-basin for many small surface drainings. The haunt of the coot and the frog is hardly the ideal birthplace of the clear, tossing brook, yet a great many streams come from just such sedgy pools.

The river's source.

The catch-basin.

The feeders of the pond are the tiny little threads of water that meet and join forces in pushing under grass and around stones, until a union of many of them makes the trickling stream. Originally these little threads are formed by the drops creeping along the seams of a rock and oozing out at the base, or they may come from the sloping surface of some ledge hidden under several feet of earth and moss. The earth and moss act as a sponge to catch the rain, which finally settling to the bottom, runs out along the bed rock. Small enough in themselves all these contributors of water taken together make the rivulet, which supplies the brook, which in turn supplies the river. The volume of the downpour is cumulative from the mountain to the shore, until at last the distillation of the hills, having passed through all its stages of life, spreads fan-like on the surface of the sea and is lost forever.

The rivulet

CHAPTER IX

STILL WATERS

Names of seas and lakes.

ALMOST any sheet of enclosed water is a sea, a lake, or a pond, as the dwellers beside it choose to name it. The nomenclature, as applied, is often very misleading. Thus, for instance, we have the term "Sea of Galilee" applied to a lake fifteen miles long by eight miles broad, whereas "Lake Superior" designates an inland sea covering an area of thirty-two thousand square miles, and having no more the characteristics of a lake than Galilee has of a sea. Any body of water, no matter whether fresh or salt, where we are at any time out of sight of land, or have a water-line for a horizon, has at least one strong feature of the ocean—immensity. The great American lakes, as we stand upon their shores, stretch out to the horizon-rim without a break, and we have small reason to suppose we are not on the New England coast looking over the Atlantic toward Europe. True enough, these great lakes have a different smell, and are not so blue as the ocean; but the beaches, the rocks, the dunes of the shore, the break of the waves, the reach of the sky, are substantially the same as those of the greater body.

Now a lake is, or at least should be considered, a body of water surrounded by land, and the name should never be applied to a body of water so large that land can from any point be lost to view. A sea is, or should be considered, another body of water surrounded by land, but so large that one does not feel its confining shores. An ocean is, or should be considered, a sea of such expanse that it is not surrounded by land, but rather seems to surround the land. The fundamental distinction here is, of course, one of size, the lake being a reservoir for a range of hills, and the ocean a reservoir for the whole earth. But this matter of size has great influence upon our tastes and preferences. We have, perhaps, a dread of the ocean, because it seems so vast and incomprehensible; but we are fond of the lake because it is small enough to be readily grasped by the imagination. The ocean in its mystery and indefinite reach has about it the breath of the sublime, but the lake in its simplicity is merely beautiful and charm-

Definitions.

Lakes vs. oceans.

ing. Again, the ocean is ever in motion. Its surface may be smooth, but there is always the heave of deep swells beneath the ebb or flow of tides, and day and night, year in and year out, it is continually beating out its surge on the shore. Not so the lake. It is ruffled only by passing storms and winds. When the winds die out it lies still in the sunlight, and not a ripple shakes its serenity.

The mountain-lake.

The small mountain-lake, shut in by shores of rock or timber, is undoubtedly the most beautiful type of the still waters. If we are on ground high enough above it to overlook the whole expanse, it will appear, when the mists are creeping over its surface in the early morning, like a mirror with breath-marks upon it. At noon, if the surface is agitated, each wave will glitter like a harlequin's spangles; and if smooth, it will reflect whatever sky is above it.

Seen from a height.

At evening it may reflect the pink and gold clouds in the zenith, and when they have burned out, it may deepen into a dark purple floor upon which the stars are spattered in golden spots. Whenever looked at from a height, it seems like some precious elixir held in an emerald chalice—a gem set in a frame of hills and forests. When we are down close to it the likeness is lost, or rather changed, for now it looks like a flat arena of blue-steel, and the tiers of hills may sweep around it like the benches of a Roman circus.

The cliff with its feet in the water, its sides dripping with the moisture of mosses and its top tufted with pines, the cave with its shadowed entrance and sunken rocks, the gorge where the brook comes into the lake, the little island, the pebbly strand, the overhanging trees and bushes, are all essentials of the mountain-lake. Even more necessary than these, perhaps, is the purity of the water—a necessity that is generally met. For though brooks may empty sand and mud into it there is no great motion of currents through the lake, and the brook water soon drops its burden to the bottom. Lake water is also, as a rule, quite clear—so clear that it will not, unless ruffled, take cognizance of a shadow, and will register sky reflection with the utmost delicacy. It may have a greenish or bluish local color, which we can see when the wind turns up its surface in little waves, and we may see this local color again at times by looking straight down into the lake depths; but there is usually no cloudiness about the water. After

Lake features.

Purity of lake water.

a storm the waters near shore may be beaten brown or yellow by the waves, and in the spring the lake may be turbid with the wash of heavy rains; but these are only temporary disturbances. It soon returns to its normal clearness and purity.

And purity combined with freshness and wildness, are the characteristics that excite an enthusiasm about lake beauty in the breast of almost everyone. We all feel it. A day spent in coasting the shores in a canoe is not only a revelation of nature but of ourselves. The drift along under the cliff, the coolness and the shade from bank and bush, the mysterious depths with sunken rocks and water-logged tree-trunks, the shoals of sand and pebbles, the little bays with pickerel grass and lily-pads, the mosses of the gorge, wake memories perhaps of an earlier, a simpler, and a nobler life. We are back to the earth again and the elements are around us. The human animal, caged in cities and taught the tricks of civilization, can never forget the nature that sent him forth.

Lake charm.

But the mountain-lake has other charms that are perhaps not so superficial or so sentimental. Its color in particular is of marvellous complexity, and if one tries to trace the cause or give the reason for this or that effect, he soon finds himself involved in many contradictions. The determination of the local hue of lake water is, to start with, a difficult task. It may be almost any color, taking its hue from the vegetable or mineral matter carried in solution. Draining a marshy or heavily wooded district, it may be brownish or amber-hued, as in many of the smaller Adirondack and Scotch lakes; if the shores are rocky, or the country drained is hard and mountainous, the hue of the water will be blue or bluish-green, as one may see in the Alpine lakes, particularly Lake Geneva. Again, in the Yellowstone region the lakes are often of varied and brilliant hues owing to the earth or minerals in the water.

Local coloring of lake water.

Vegetable and mineral colors.

But the actual color of the water when taken up in a vessel, and the apparent color lying in the bed of the lake, are two different things. Local color, especially if it be delicate, is influenced, changed, oftentimes utterly destroyed so far as our vision is concerned, by backgrounds and reflections. For instance, the bed or bottom of a lake, where the water is shallow, may decide the apparent hue, just as the green

Colors of backgrounds.

grass may decide the coloring of a shadow falling upon it. Pure water is in itself a most exquisite and subtle blue, but spread over a bed of yellow sand under sunlight it may appear yellow or perhaps greenish in hue. Even when the water is too deep to see the bottom, the latter may have some determining influence on the color by mingling with or illuminating it.

Local hue and reflection.

Reflection is, however, the more powerful factor in leading us astray as regards local hue. Smooth water, like a mirror, is always throwing back from its face some likeness in light, form, and color of whatever happens to be above it, be it rock, tree, bank, or sky. The water may be green in, let us say Lake Placid, but the reflection of the clear sky from its face makes it appear blue. Even when the surface is agitated and the reflection is broken, there is always more or less flashing light from the sky along the tiny facets on the backs of the waves. We can only get the local hue by shutting out the reflected hue. A gray sky with a ruffled surface dispels or breaks reflection sufficiently to give us some notion of local color; and it is during rain-storms and squally weather on sea, lake, and river that we gain the truest knowledge of the actual colors of waters. Sometimes, under lake-banks or under the shade of an overhanging tree that completely shuts out the sky, we can see the hue of the water; but here again we have to reckon with the reflected green of the tree or the color of the bottom. We must look *through* the reflection, and not *at* it. Even then, and under the most favorable conditions, we are often deceived into thinking the water one color, when in reality it is another. And just here begins another complication.

Confusion of hues.

I said that smooth water, like a mirror, is always throwing back from its face " some likeness" in light, form, and color of whatever is above it. Its light is always feebler than the original, but its color is almost an exact likeness, *provided the water is very clear and pure.* The tall cumulus cloud, the blue sky, the dawn, the sunset, and the rainbow, are all given in the lake reflection with accuracy, but with perhaps more delicacy than in the original. This is particularly true of the sky and the clouds. The hue in the reflection is more refined and silk-tissued, and the lines of the clouds are less positively defined. Now deeply colored or darkened water will reflect a likeness, too, reflect it in form quite as sharply as clear water, but the color will also be deepened and darkened. On

Phases of reflection.

a peat-water lake, like Loch Laggan in Scotland, the white clouds show gray, the gray clouds look sooty or smoky, and the cerulean blue of the sky turns to deep ultramarine. On Loch Dhu, a dark little lake in the Grampians, surrounded by high hills covered with nothing but bluish stone and yellow-green grass, one can see in the wave reflections, the grass turned to dark orange and the stones to cobalt-blue. Dark local tones in the water will darken the colors in reflection, and light tones will lighten them. A muddy or yellow lake will not reflect a brilliant hue of any kind without bleaching it. As for waters neither light nor dark, but nevertheless positive in hue, they will often tinge the reflection with their own intensity. Thus the waters in the Venetian canals reflect the side of a black gondola, but the reflection is not black; it is greenish—the local color of Venetian water. Again in determining hues, local or otherwise, much depends upon the angle from which lights, colors, and reflections are seen. From one point of view the lake may be all local color; from another point of view it may be all sky reflection. So that when the disturbing elements taken together are considered, the problem for deter-

On dark waters.

On strong-hued waters.

The point of view.

mination will prove anything but easy. But I must mention just one more complication that should be simple of solution, and yet is not always found so.

In studying effects on the water we are prone to confuse shadows with reflections. They are two separate things, though in effect they may sometimes be merged into one. That is to say, a tree may cast its reflection in the water and its shadow on the bank ; but if the sun is just right, both the shadow and the reflection may fall in the water, as in the case of an overhanging bough or the arch of a bridge. The shadow in such cases is usually absorbed by the reflection. Shadows upon water are usually very feeble, and where the water is *deep and perfectly clear* they are hardly noticeable at all. If the water is shallow or muddy, the shadow is stronger, because it has some background to fall upon ; but even then it is not so strong as when falling upon ground or grass. On deep water the shadow is seen as a thin, smoky form upon the surface, whereas the reflection is seen receding into the depths. And at certain angles the shadow does not appear at all. If one is standing on the bank of a pool with a small tree beside him and the sun is be-

Shadows vs reflections.

Surface appearances.

hind him, he will see his own shadow as well as that of the tree cast upon the water, but he will see no reflection. His friend, standing on the opposite bank of the pool and looking toward him, will see no shadows, but in their place the reflections of man and tree. A canal in Amsterdam, with houses and trees on either bank, will often mingle shadows and reflections, but to us it is merely a case of shadow on one side and reflection on the opposite side. If we stand in the centre of a bridge and look up the canal, we shall see little of either houses or trees in the water; we shall see only the long reflection of the sky.

On the lake the strongest reflections are always to be found under some overhanging bank or in the shadow of some thick-leaved tree; and the darkest reflections of all are seen at night when the only illumination upon the water comes from the sky and the stars. Very beautiful are these night reflections seen from a boat. Years ago, when the lakes and streams of Minnesota were in their prime, and the great elms arched the sloughs winding from lake to lake, the canoe trail at night was picked out by the small spots of sky showing on the water like loopholes in the vast density of reflected foliage. As the canoe slipped over these sky-lighted spots, the stars could be seen trembling for a moment in the water, and then the sweep of the paddle would scatter them into a thousand tiny flashings. A similar effect can be seen almost any night on the mountain-lake where trees or banks overhang the water.

It is, perhaps, necessary to explain still further the statement that water is always throwing back from its surface "some likeness" of whatever is above it. The likeness is not always and invariably exact in form, any more than it is in color or light. In the first place, the reflection is always the reverse of the original, as is a human face in a mirror. That is to say, left is right, and right is left. Secondly, the background of the reflection may be different from the original. Standing on a bank twenty feet high and looking across water fifty feet to a low shore, we may discover that a bush overhanging the water has a green meadow for a background, but in its reflected image it has a blue sky for a background. Thirdly, the tint or shade of this same bush in the reflection is not the tint or shade of the original; and this for another cause than local color in the water.

Variations in reflection.

The likeness inexact.

We are looking down upon the bush and see sunlight upon its green leaves; but the reflection shows us that under-portion of the bush which is in shadow.

The rule governing our perception of reflections is a familiar one: The angle of reflection is always equal to the angle of incidence. Practically applied to our illustration, this means that standing twenty feet above the bush and fifty feet back from it, we see in reflection just what we should see in the original did we stand twenty feet *below* the bush and fifty feet back from it. Again, if standing twenty feet above the surface we can see a portion of a mountain-peak reflected in the water, then we could see just that much of the peak itself if twenty feet below the surface. Every Swiss tourist has seen Mt. Blanc mirrored in the Lake of Geneva, though the two are some forty miles apart. The reason is that Mt. Blanc is some three miles high. By increasing our height we see less of the reflection in proportion, as by increasing our depth we see less of the original. At the bottom of a well, looking up, we should be able to see only the sides of the well and the sky; if at the top of the well, looking down, we should see the same things in reflection.

The angle of reflection.

Anything that disturbs the smoothness of the water also disturbs the clearness of the reflection. A breeze may cause it to disappear, and at the same time may make the shadow more apparent. Rough water will show the shadow of a flying cloud almost as clearly as a field of grain or a hill-side meadow, without giving us more than a hint of the cloud's reflection. When water is slightly ruffled the reflection does not disappear at once, but is lightened or silvered in color, and at the same time it is elongated. This everyone must have observed at night, with the moonlight falling on the lake. When the surface is perfectly calm the moon shows in the water like a round ball; but as soon as the surface is ruffled we have the elongation of the reflection into that flickering trail of light called the Angels' Pathway. Elongation is seen again in the reflection of the artificial lights in a harbor, where the water is always disturbed. The reflection writhes and twists in the water like a fiery snake. The sails of ships, houses, trees, almost any object near or upon the water, will show elongation in the same way. Everyone who has been in Venice knows the effect of the orange sails reflected in the swaying waters; and the long wriggle of the blue-and-white gon-

Elongated reflections.

The Angels' Pathway.

dola-posts in the ever-agitated Grand Canal is matter of common observation.

Lake sentiment. In its reflections, shadows, lights, colors, forms, there is nothing in nature superior to the clear mountain-lake. It has no sentiment, no feeling whatever, though we often speak of it as though it had; but there is no limit to the emotion it can arouse in the breast of humanity. I am not privileged to speak of this at any length, for I have set myself the task of writing about nature as it is, rather than about the romance it can create; yet, no one can be insensible to that romance. The splendor of early morning on the lake, the fresh breeze, the waves dashed back by the bow of the canoe, the glitter of myriad points of sunlight, the blue sky, the voyaging clouds, the sentinel mountains that stand like giants around the little basin, are all productive of impulsive feeling. Nor can anyone be quite indifferent to the silence of those mountains at night, the slow rock of the lake *Moonlight on the lake.* waters, the shimmer of the stars, and the moonlight weaving a pathway of splendor from shore to shore. Beautiful in themselves, and for themselves, these features are not the less potent in awakening thoughts of beauty in the mind of man.

The sentiment is, of course, wholly of human origin; and that part of it which relates to the weal or woe of past humanity is not with us here in America. The legend and the story cling about European lakes and make them romantic; ours have only their material beauty combined with a dash of untarnished freshness that belongs to an unworn world. But that material beauty is quite sufficient in itself. Without pride of place or breath of patriotism, the American may venture to think that such waters as Lake George are not outranked in beauty by any lake waters on the face of the globe. To be sure, the Swiss lakes come in and claim high place in any such comparison. The Lake of Lucerne has great charm as well as great beauty about it, though, perhaps, it is a little dwarfed and obscured by its high mountains; and surely the Italian lakes are exceptionally fair and lovely to look upon. The Irish and the Scotch lakes, too, are famed for their beautiful borders and graceful forms, though in purity of lake color they cannot rival the waters of Geneva or Como.

But, again, I come back to query: What is so fair as Lake George? It has all the marks of natural beauty unblemished by cities and

Material beauty.

European lakes.

Lake George.

artificial growths. A sheet of clear water in a framing of green hills, dotted by many lovely islands and colored by as bright a sky as ever arched the earth, it seems to epitomize all lake loveliness, and to exemplify the luxuriant splendor of untrammelled nature. The breath of the wilderness is still there, though man has begun to tenant its shores in places. The wind that blows over it is pure, and those timbered heights above it are, as yet, comparatively untrodden. Its beauties seem as bright as when the earth and the firmament and the sea were first created; and to-day, as for many centuries, a light seems to come out of the west at sunset, tingeing the green-garmented shoulders of Black Mountain with a golden hue unknown to the Alps and the Pyrenees—a hue belonging to the primitive world, put on by nature for its own splendor and its own pleasure.

The typical mountain-lake.

A pond or a pool is often little more than a diminutive lake, filling a depression and produced by an embankment after the fashion of almost all still waters. It differs from a mountain-lake by usually having low-lying shores without tall timber or rocks, a sandy or muddy bottom, and perhaps, flags, rushes, and rough grasses growing along its shallow

The pond.

margins. Almost every town has a local body of water of this description which answers to the adjective of "Silver," "Blue," "Fresh," or "White." The sarcasm of the name is unconscious but not the less biting, for the pond is generally a stagnant, malarious little place, with the frog, the bull-head, and the snake for occupants—its waters yellow and its shores green with scum and parasitical vegetation. Its principal charm lies in what it may reflect of light and color from the sky.

Quite different from this is the pond that lies away from civilization, hidden, perhaps, in the depths of some forest where tall trees come close down to the shore and peer into the water, where the vines and underbrush make an almost inaccessible bank, and where the brown water, lying over sunken trees and beds of leaves, makes a dark mirror for the sky. The silence, the solitude, the utter isolation of the woodland lake seem to give it interest. So, too, with the prairie pond, lying out on the treeless plains in its fringe of wild rice—the spot where once the swan and the wild goose paused in their migratory flights, where once the buffalo came to wallow, and the Indian and his pony to drink. Birds and

The forest pond.

The prairie pond.

beasts and Indians have about departed, but the prairie pond in its wild rice circlet still exists; at morning and evening the red of the sky, the pale yellow of the rice, the green of the flag gleam upon its waters; and at night the moon and the stars are reflected from its shining surface. It seems about the only surviving feature of a nature that has rapidly passed away before the axe and the plough. It belonged to the Indians, and is associated with them. I can see them now, a band of fifty or more, bonneted and painted for war, dashing down a divide and plunging into that prairie pond to let their hard-ridden ponies drink. They pause for only a moment, the ponies pushing their noses deep under the water, and then, at a signal yell they come rushing out of the pond, through the rice, through the tall prairie grass, and vanish like dusky spectres over the next divide. They come and go no more. The prairie grass has turned into a wheat field, and the prairie pond is the watering-place for herds of cattle.

In Indian days.

Artificial waters.

Almost any little pond or basin of water adds to the interest of the landscape, however humble or even mean it may be intrinsically. It is always a bright surface and can reflect beautiful coloring and light though it have neither in itself. Even artificial waters, though they are usually dull and lifeless in body, are better than none at all. The formal beauty of the landscape-gardener is about them, but taken in connection with houses, trees, and skies, they may have a certain artistic charm. This charm is well shown in the pleasure-lakes of various European estates, and particularly in the canals of Venice. The canals were originally the natural tide-ways between islands, and when the city was built the mud-banks formed the foundations for the houses, and the canals themselves became the water-streets of to-day. Not a place in Europe can show such beautiful and picturesque compositions as Venice. The color, light, and reflection of the city and its waters are world-famed. The Ducal Palace, St. Mark's, the towers and domes and palaces that heave out of the blue-green tide, change their color fifty times a day with the changing of the sky; the swaying waters of the canals are tremulous with direct and reflected light; and the ships, sails, wharves, and bridges splash the horizon-line with countless patches of orange, blue, red, and yellow. And these are only the pronounced hues. From

Venetian lagoons.

The canals of Venice.

every broken wall and water-worn step, from post and stunted tree and marshy shore, are thrown off those indescribable tints that seem always identified with decay. Everything about Venice seems to reek with color. It is the hectic flush of the dying. But how very beautiful it is!

Holland canals.

The canals of Holland are quite as artificial as those of Venice, but they are different in appearance. They have a more even surface, little or no motion, and are often foul in their stagnation. Nor has the local color of the water the life about it of the Venetian blue-green. It is dull, dark, often brownish in hue, and perhaps for that reason makes an excellent reflector, throwing back the houses, the trees, the great white clouds, and the blue sky with superb effect. Again, the Holland canal landscapes in their arrangement are not so varied as those of Venice, and the waters of the back country are quite different from those in Amsterdam.

The country canals.

The country canals, with their low banks and their rows of willows, the slow-moving boats with lazy sails, the ditched meadowlands, the groups of cattle, the long-armed windmills, lend to a quiet pastoral effect and make Holland one of the most restful places in the world. It does not startle or oppress one like a mountainous country, but is ever quiet and peaceful, having about it the serenity of its smooth-faced waters.

But these waters of Holland and Venice, with all their charm, have really little of nature about them; or, at the least, what there is of nature is so alloyed with the artificial that we think of them only in connection with humanity. After seeing them we instinctively hark back to the mountain-lake. It seems to lie so much nearer to nature's heart. Its shores and islands, its water and sky, its lapping waves and fresh-blowing winds, are stimulating, invigorating, strong with the strength of youth and instinct with life. Beautiful in repose, the mountain-lake is not without beauty when agitated. Even in storm, when the first heavy drops of rain spatter the smooth surface and the sweep of the wind may be seen in the ruffled water-line, when the waves are dashing and tossing on the island shores and the roar of the tempest can be heard along the sides of the wooded mountains, even then the mountain-lake is more beautiful than almost any other body of water in repose. After many summers spent at Venice, I hope I am not insensible

The mountain-lake again.

Lake beauty.

to its splendid sea-flooring, and all the crumbling glory that sags above it and is reflected in it, but as an example of nature's beauties it is hardly admissible. The mountain-lake is nature—pure, simple, and undefiled. No one can fail to admire it and love it. It is one of nature's brightest jewels set in her green girdle of hills.

CHAPTER X

THE EARTH FRAME

It was the teaching of our childhood that, if we divided the earth into land and water, and allotted one-quarter to the former and three-quarters to the latter, we should have the proportionate distribution of these elements. Too many of us, perhaps, accepted the statement literally, and when we looked upon the map, vaguely wondered if the water floated the earth or the earth the water. Even in maturer years it is not easy to realize that all the water is on the surface, that the earth's hollows and depressions hold it as in cups, and that after all, it is the earth and not the sea that is the dominant body. The volume of the sea is enormous, to be sure, and that of the air is still more so; yet the land is not moved by them, but is the mover. It is the central body around which air and moisture gather—the solid upon which these surface-coverings rest.

Earth and sea.

The form of the earth is, of course, globe-

The earth's surface.

like, and if we could see it from a distance denuded of atmosphere, it would doubtless appear as a ball of water with patches of land like large islands projecting above the surface. The greater part of the surface is on one level — that is to say, sea-level, which we accept as the standard. Above it there are elevations of land in points and ridges called mountains projecting upward some thirty thousand feet; and below it there are depressions or holes in the sea extending down some thirty thousand feet. There are numerous exceptions to the rule of the sea-level marking land above and water below. Some of the plains and basins are below or above the sea-line. The margin of the Dead Sea lies thirteen hundred feet lower than that of the Mediterranean, and the

Inequalities of surface.

great lake of Titicaca in the Andes, with an area of three thousand square miles, is nearly thirteen thousand feet above the Pacific. From Titicaca downward there are hundreds of bodies of fresh water conserved in great basins of the earth, which are as upland reservoirs to the sea itself.

It is fortunate, indeed, that the earth has these upland reservoirs, fortunate that they are so equally distributed over the face of the earth. If it were not for them there would be more Saharas of desolation, and the green of the earth in many a tract now lovely in light and color, would take wings and vanish into space. It is color that gives the glow of life to the earth, and yet this great beauty might disappear without weakening in any way the frame or form of the globe. The earth could and would exist, and swing on in its orbit, were there no life, no light, no color upon it. The modelling of the mountains, the deep-incised river valleys, the flat spaces of the plains, the hollow depths of the ocean-beds, would remain substantially as to-day. For the material of which the earth is formed is so cohesive that its shape would probably hold for centuries after its seas had evaporated and its atmosphere had passed away. The interior of the globe may be fire or rolling vapors, or simply solid matter; man speculates about it without absolute knowledge. But about the crust he is quite certain that it is rock formation, made in different ways and at different geological periods of the earth's history.

Doubtless, the whole globe was at one time covered with sea, and when the waters receded from the table-lands there was a hardening of the sedimental deposits; doubtless, there were at other times great streams of hot lava forced up from below by volcanoes and spread over vast surfaces, cementing the sedimental deposits into rock strata; and doubtless, again, chemical action and change by and through air, water, fire, produced other rock masses with which geology acquaints us. The forms of rocks, their twisted, broken, or waving strata, were caused by convulsions of the earth (either expansions or contractions of the crust) which, following the form of a wrinkle or a fold, have heaved the surface in some places and depressed it in other places. The deposits which we call soil, together with the bowlders and loose stones, are but the grit from rock formations, broken away by frost, wind, and rain, and washed down into the valleys by the brooks and streams.

Formation of the crust.

The human being has always had a very keen appreciation of the earth's volume and substance. Earthquakes may shake his house, but not his faith. The tremor is but temporary; he still believes in the solidity of the earth under his feet. And yet how seldom he thinks of the immensity of the structure, its continuity, its long endurance, factors which have made possible its

Solidity of the earth.

cohesiveness and its solidity. But a few years ago one could ride over the prairies of Northwest America—could ride for weeks up and over the rolling divides, through the tall grass where the horse's hoofs made scarcely a sound, where there was no tree nor lake nor river nor any trace of human habitation. There seemed no end to the vast stretch of grass and sky. How very impressive it all seemed! How calm and serene the great motionless swells of the prairie! Rolled in their wave-forms, they had not moved nor changed. They were probably cast in those forms ages before the Indian and the buffalo came. The tall grass wove a protecting mesh over them so that wind and rain should not shift them. They lay silent and immovable for so many centuries. But the plough is now ribbing their hollows and breaking their backs. They will wash into lakes and rivers and flatten down into plains, now that the white man and his civilization have come. *Permanence of the prairies.*

But a few years ago one could hunt the deer, the bear, and the moose in the great forests of Minnesota and Wisconsin—could hunt and walk for days and weeks without coming to the end of those "Big Woods," as they used to be called. The great pines, oaks, sycamores, and elms, *In the forest primeval.*

through whose tops had whistled the winds of so many winters, how sturdy they stood! The fallen giants of the wood lying prone upon their faces, blown down years and years ago, looked sound and substantial under their moss coverings; but the pressure of the foot would show that they were dust—a semblance merely of form. The scattered leaves and pine-needles seemed a very thin earth-covering, but one could dig deep and still turn up the crumbled mould of trees. That forest must have been before ever the hosts of Ur or Assur were brought forth. Here it stood, its trees holding in solid ranks, the older dying off, the younger springing up to take the vacant places; yet apparently the forest never shifting, never changing. Scarred it was in places by fire and windfall, but these were mere spots that in no way impaired its calmness, serenity, and appalling majesty. It is all but gone now, yet the destruction was not nature's own. The axe has laid it low, the rivers have carried down the logs, and man has sawn them into lumber and shipped them around the world. The forerunner of civilization is destruction, and its follower is always desolation.

But Sahara is still the same Sahara that

Age of the "Big Woods."

Menes knew; it at least has remained quite undisturbed. The traveller may strike off from the Nile and ride—ride west for days without a change. There is with him always the glaring sunlight, the sand and rock, the torn and ragged wady, the star-like glance of light from quartz and mica, and overhead the rose-hued sky. Nothing but barren waste below and burning heat above. The two expanses circle and enclose him as he stands upon his central point of sun-fire. One may ride on for hundreds of miles and still there is no change. The opal flash of sands, the glaring rocks, and the trembling, heated atmosphere—that is all. How silent and motionless the vast desert! Simoons may blow and drift the sands hither and thither, but the general appearance does not alter. It never alters. The desert steeds of the Pharaohs perished in these wastes ages ago, as yesterday the caravan of the Mecca pilgrims. The Sphinx with its face to the sun and its back to the desert, has felt the far-travelling waves of sand lapping its shoulders through no one knows how many centuries of desolation, but the sands were there before ever it was carved. Will they always remain as now? Who knows what changes the engines of civilization may

The changeless desert.

The sands of Sahara.

work? The northern Libyan desert may yet form the bed of a great inland sea.

And is there nothing more permanent about the earth than prairie grass, and forest trees, and shifting desert sands—nothing more substantial than these? If one stands on the height of Mürren and looks across to the base of the Jungfrau, he may think differently. What a stupendous pedestal for that white-capped young goddess of the Oberland! The wall of rock is simply tremendous in volume. It stretches wide, it reaches high. It is the vaulting of the globe—a glimpse of the understructure of the crust—exposed to view by the accident of a valley. It is this massive vaulting that apparently holds the globe together as the shell does the egg. It stretches around the whole earth; and the forests, the sands, the mountains, the seas, are related to it only as the mosses, the wind-blown dust, and the rain-pools to the Coliseum's walls upon which they lie. The structure is almost stifling to the imagination, so great is the plan, so small is the mind of man to grasp it. If we look away from the wall of rock up into the far valleys, where the blue air lies packed in between the uplifted peaks, and listen a moment, we shall realize a silence so in-

The vaulting of the globe.

The base of Jungfrau.

tense that it can be heard. The very air is freighted with the hush of a majestic presence. It comes to us with an indescribable vibration which has about it the hum of distance and immensity. The sea-shell which the child holds to its ear suggests the same wondering tale. It is something that tells of power and eternity.

How many centuries that masonry has stood ! How many centuries, through heat and cold and earthquake shocks, it will continue to stand ! *Understructure of the mountains.*

> "When you and I behind the veil have passed,
> Oh but the long, long years the world will last."

Surely, it has a strong foundation. Not here in the Alps alone, but in the Caucasus, the Andes, the Rockies, its supporting courses of rock keep peeping out. It was only geological yesterday that these breaks in the crust became apparent; and, perhaps, under the rounded, timber-grown slopes of the Appalachians, or the old, old, grass-grown plains there rest still mightier strata of rock. Again, the scratches made in the great shell by eroding rivers like the Colorado, the Mississippi, the Hudson, the Rhine, the Danube, are but recent gnawings of water upon rocks. Under the table-lands and the foot-hills may rest an untouched rock-structure *Hardness of the rocks.*

of sterner stuff. How like steel or flint that base supporting the Jungfrau! Those who cut the tunnels through Mt. Cenis and St. Gothard found out how compactly that wall of the Alps was put together, and of what rock quality it was built. It would seem as though those strata were laid, one upon another, with the aim and the design of their enduring forever.

Nature's building principle.

And how could the frame itself have been planned better? Arched at every point by the great rock-beds of plain or mountain, it is more cohesive than any dome of human masonry, be it of the Pantheon, or of Hagia Sophia, or of the Taj Mehal. The architectural drum is but a half-globe placed like an inverted cup upon supporting walls and members, but the earth is as complete in its rotundity as in its continuity. Braced by its own curve, the atmospheric pressure from without has as little power to crush it in as the possible gases and vapors from within to bulge it out. Doubtless, ages ago, when the earth was

The self-supporting globe.

soft and pliable, its whirling motion through space made it round, much as the rain-drop rounds itself by passing through the air; and now that it has hardened, it is not likely to lose its shape. All worlds are round. Nature is intent upon building for eternity, and it chooses the strongest building principle of all—the self-supporting globe.

We cannot see upon the globe itself the great, sweeping lines that make the earth circle, but, as intimated some chapters back, we may gain some suggestion of them from certain sky curves. The upward and inward arch of the blue sky, the visible envelope of the earth, but repeats the curve of the earth itself. Nothing gives the feeling of the globe's rotundity so effectively as this; and yet I have been told by many observant people that they could not see this curve of the heavens, that the blue looked to them perfectly flat, and that the monotone of the color, the brilliancy of the light, destroyed all sense of form. Perhaps the curve would be apparent to them if they saw it cast upon the sky by the earth's shadow instead of suggested in the illuminated blue. This can be seen almost any summer evening, when the sky is free of clouds and the sun has just fallen below the horizon. Facing to the north and looking at the zenith, we are aware of the light slowly fading out of the west. Everyone sees and knows that fading light, but few of us ever

Earth lines

notice the coming shadow which follows after. I do not mean the darkening of the hills and valleys and waters about us, but the shadow on the sky—the great earth shadow stealing up toward the zenith from the east. As the light passes down the vast incline and below the western horizon, this shadow coming up from the eastern horizon, moves slowly into its place. It is not readily seen at first, but after a few observations the eye becomes more quick to note its presence than the mind to conceive of its vastness. It is in this shadow, drawing up and over the sky like a thin veil, that one can often see the suggested curve of the earth. It is at times very obvious, but it never seems so clear and pure as the curve of the blue sky in the morning, because it is frequently confused with lower shadows. Nothing, indeed, can excel the marvellous sweep up and over of the illuminated blue. The dawn-light mounting the sky does not go beyond it, and the noblest spring of bridge or dome designed by man looks strained beside it. It is drawn so perfect, and it rests so serene in its perfection, that even the arch of the rainbow seems almost like a child's toy in comparison with it.

It has already been suggested that a glimpse

The earth's shadow on the sky.

The arch of the sky.

of another earth line is afforded us by the horizon. This horizon curve can be seen to the best advantage from the cross-trees of a ship in mid-ocean. There the circle that sweeps about one is quite complete; and the line one sees is the edge where the world slips down beyond our vision. Again, how perfectly that curve is drawn; and on a clear day how embracing is its sweep! A similar, but perhaps not such a perfect, effect can be seen on the alkaline plains looking out from some tall butte across the lowly buffalo grass, with the wavering heat of the plains rising upward instead of the ocean moisture. It is a smaller circle, a smaller earth line, that is thus revealed to us; but what a hint it gives us of the greater lines which must be merely its enlargement! *Horizon lines.*

The demarkations of light and dark against the sky are about the only glimpses of the earth-curves that are vouchsafed to us; but the principle of rotundity—the curved line so often called "the line of beauty"—is shown to us in almost all the earth formations. The zoöphite, that builds a rounded cell in the rounded coral, making by aggregation the round island in the Southern seas, is typical of all creation. The law of the circle that curves the waves of light *The curved line.*

and color on the interior of the sea-shell also curves the prairie, arches the hill, rounds the lake, and bends the river. The line forever sweeps in and around. Even when there are apparent exceptions in nature's products, as in the forms of crystals (and there are round crystals, too), or the sharp needles of a mountain-peak, there is an attempt to amend the fault, as it were. The winds, the rains, the frost, the heat, immediately set about rounding and curving the knife-like edges; and the peak, which at a distance looks sharp and angular, proves to be round and smooth when seen close to view. The hill, that to-day is dome-crowned like a Byzantine lantern, was once snapped and splintered upward by the sharp fold of the rock strata; and many a coast-lying island, that now is carpeted with soft rolls of green sward, was originally banked up in a ragged heap and pushed ahead of a glacier in the great Ice Age.

The law of the circle.

The circle is indeed nature's great working principle. Organic and inorganic matter—winds, storms, clouds, tides—all display it. The life that springs up from the earth withers and returns to the earth again, rising and falling in the lines of a jetting fountain; and the

More circles.

evaporation from the sea, that carries inland and descends as rain, filling the lakes and the rivers and finding its way back to the sea, again exemplifies the circle. All the elements— fire, water, air — show it. Yes, even the planetary system pays allegiance to it; and as the moon circles the earth, so the earth circles the sun, and the sun itself with all its planets, is following a mightier curve about some greater orb, or is, perhaps, drawn inward by the sweet influences of the Pleiades.

And is the physical or intellectual life of man any exception to the rule? Tribes travel from east to west for ten thousand years, following the track of the sun, until at last they rediscover the cradle of the race—reach the spot from which migration first started; minds build up thought for ages, advancing, as they suppose, until at last they find they are but rediscovering the truths known to the ancients. They have completed the circle, and the circumference is limited. Out of it and beyond it man cannot go. Once, in an Eastern church, my attention was attracted by a beetle in the dome wheeling and beating against the encircling stones, vainly trying to get out. Around and around that dome, hour

Human circles.

The limited circumference.

Mental and physical limitations.

after hour, he droned his way; but the encompassing arch offered no exit. Ah! how the blue arch of heaven closes in upon the winged flights of the Buddhas, the Mahomets, and the Platos! They may drone on in circles beneath it for ages, but the human mind will never break through and beyond it. Man is not different from the other creations of nature. His lines are cast in curves, and he glides along them from the cradle to the grave, unconsciously obeying the universal law. He thinks his movement progress, but is it anything more than change? His is a different segment from the one his father traced, but is it not a part of the same circle? Nature never designed that the human should be exempt from the universal law. The circle binds him as the world spins. His deeds and his thoughts stretch upward, as flowers spring aspiringly toward the sun, but ever and always they are curved in and turned back upon themselves, sinking into the earth mould from whence they rose.

CHAPTER XI

MOUNTAINS AND HILLS

THE mountains have more than once been characterized as the "backbone" of the globe or of the continent, but one cannot think the simile other than misleading. The globe has no more backbone than the sun-baked bowl of a Zuñi Indian. It has not even a rib or a vertebra; and the mountain-ridge is no more its binding member than the upheaved track of a mole across a garden is the band that holds the garden together. The mountain-ridge, however, is not produced in the same way as the mole-ridge. The great layers of rock, piled up on end like the poles of an Indian's tepee, that make the Alpine peaks, are more the result of lateral pressure than direct upheaval. They were pushed up as a wrinkle in the crust of the earth, and the beds of loose soil that lay above the rock were rolled back into the valley, leaving the ragged edges of the crust exposed to view. In other words, a mountain-

Mountain ridges.

ridge is a bulge or break in the dome of masonry, and not a string-course where an extra layer of rock is placed for ornament or strength.

How mountains are formed.

It seems to be the present scientific conclusion that mountains are not formed so much by volcanic action as by the folds or laps in the crust made by the contraction of the earth as it grows older and colder. The illustration used is that of the skin or surface of an apple. It wrinkles in folds as the apple withers and decreases in size; and these folds in the skin of the apple correspond to the mountains and valleys of our earth. The illustration and the conclusion are both very plausible. The long line of the Rockies once lay, perhaps, thousands of feet beneath the flat bed of an inland sea, but some contraction of the earth, some great sinking-in of the crust on either side, caused a corresponding fold to rise, and the result was the long range of mountains from Alaska to Patagonia. The high point of the fold came just on the central line of the ridge, and from that outward, on either side, this fold was less marked, producing near at hand the smaller spurs, then the slightly heaved foot-hills, finally merging into the undisturbed plains. The Alps were doubtless formed in a similar

The Alps.

manner, except that the bulge or uplift was more abrupt from being localized within a comparatively small area.

It is popularly supposed that the Alps, the Himalayas, the Rockies, are the oldest and the most permanent of the earth's formations; and the "steadfast mountains," the "everlasting hills," the "eternal Alps," are the common figures of speech used about them. But the lofty mountain would seem the youngest of the earth's formations, and so far from being "eternal" or "everlasting," it is wearing away much faster than the lower heights. For the mountain is an exposed point of land—a high point—and is always suffering from the wear of the elements. Of these elements, water is the most destructive of all. The snow-cap of the peak is a condenser and a cloud-maker for the vapors of the plains. It rains or snows on the upper ridges night after night when never a drop or flake falls in the valley. The water collects in swift-running streams, the more destructive for their velocity, that cut and rib the mountain-side. The soft portions of earth and rock are eaten out first, and the hard parts crumble from lack of support. Then the sun-heat expands, the cold contracts and splits, the winds and rains erode, the

The age of mountains.

Agents of denudation.

glacier grinds, and the avalanche tears. All told, the wearing-down process is very effective. Perhaps, long before the Alps were bent upward, the Appalachians were towering in the air, the loftiest mountains on the globe; but countless ages have given the elements the chance to wear them away, until to-day the ridge lines are almost horizontal, and the once spectral peaks, that may have projected like dragons' teeth, are no more. The torn and splintered look of the Alps and the Andes but prove their youth. The time will come when they will be worn away to low hills, and finally reduced to the common level. The flat plains, which we never think of as rock-based, are perhaps, in their foundations, the spots of earth that have remained undisturbed the longest of all.

Many of the hill-ranges that lie about us to-day are merely very ancient mountains denuded of their mountain features by the constant wear of the elements. But all of them are not of this character. Some of the hills were originally formed not by a fold but by a crack or split in the crust which has allowed one side to sink down and left the other side exposed to view, in abrupt cleavage, as it were. Such hills are not usually very high, and their tops are flat, without peaks. Many of the round mounds that are to be seen on the plains and watersheds of the world were probably formed in still another way. They are composed of *débris* of clay and gravel, and were perhaps pushed to their places by some glacier of the Ice Age. They were not caused by breaks in the crust, and have no splintered rock strata about them. Then, again, there are so-called hills that are not hills at all, but exposed portions of the earth's crust caused by the erosion of rivers. The bluffs that fringe the banks of the Upper Mississippi are not mountain-heights; they indicate merely the level of the prairie. The river passing over a sand-stone crust has cut through it and sunk its bed five hundred feet or more below the prairie surface. Many a hill or mountain in the valley of the Hudson or the Connecticut has been formed by the water passing around it and wearing through the softer portion of the rock, leaving the harder portion standing. It is even said that parts of the Catskills, and many of the mountains in Colorado, were formed, not by folds in the crust, but by erosion—the cutting out of the valleys about them by water.

It is seldom that a mountain-ridge or chain

Exposed crust.

Mountains by erosion.

The approaches to mountains.

rises abruptly from the surrounding country. The fold has usually left its mark for hundreds of miles on either side of the chain, and the ascent to the topmost peaks is made by a gradual rise from the plains to the table-lands, and from these to the foot-hills, so that frequently the mountain-climber finds himself thousands of feet above sea-level before the outlines of the ridge appear at all. This is not, of course, true of the Alps, where the deep valleys enable one to come to the base of Mt. Blanc, for instance, and see the mountain itself towering twelve thousand feet higher up; but it is quite true of the Rocky Mountains, especially in the Montana region. The ascent is gradual from the prairie "divides," which one thinks of (erroneously, no doubt) as the little wrinkles of the earth's surface, through table-lands and foot-hills covered with vegetation and

Seen from a distance.

cut by beautiful valleys. The hills near at hand are bright green, but they grow bluer and the valley shadows paler as they recede from us, and oftentimes in clear weather one can see far away, beyond the timber-crowned slopes of the foot-hills, the faint gray silhouette of the high mountain-ridge, almost lost in the blue of the sky. The fold of the earth crust is usually a long one.

In considering mountains for their picturesque appearances, the ascent of them claims some attention, for, oddly enough, mankind in general will have it that the only way to see a mountain or a valley is from the mountain's top. One marvels at the universal predilection for the "view," and at times the wonder grows if the energy spent in scrambling up to high places is not worthy of a better cause. It is all of a piece with hanging over Niagara and being agitated by the "bigness" of things, or looking through the reverse end of the opera-glass and wondering over the smallness of things. The man in Paris who climbs the stairs of that wearisome Column of July, to see the city lying below him like a checker-board, is cousin-german to the man who climbs Mt. Blanc to see the "view," and, incidentally, the smallness of his Chamonix hotel lying below him in the valley. Like other people, I have done my share of mountain-climbing, but I never felt repaid for the exertion, and I may add that I never had much sympathy with the "view" as seen from mountain-tops. It is usually said to be "grand," but to me it has been so only in a scenic, panoramic way. Even from such comparatively low places as the Catskill Mountains, or the

Mississippi bluffs, or the Leopoldsberg near Vienna, the great expanse of territory to be seen in the vista looks "mappy," and I cannot imagine anything more dreary than to live upon such heights, straining one's eyes and imagination over the lines of a river-valley, with its dotted farms, towns, lakes, and woodlands.

From the high Alps.

Doubtless, I am lacking in appreciation just here, and yet I must add further that the "view" from the high Alps is even more depressing and unsatisfactory to me. The helter-skelter confusion of snow-fields, great glaciers, gray needles of rock, and flashing blinding light may be sublime in the sense that chaos is sometimes sublime, but it is hardly beautiful. If one looks about him the masses are too big for comprehension, the eyes grow weary looking at them, and finally the imagination—the power to conceive the scene—breaks down. If one looks over into the valley it is the world seen through the small end of the opera-glass again—the scale is too petty, too map-like. In fact, the "view" from the mountain is something more than the unusual produced by distance; it is in measure a positive distortion so far as our eyes are concerned—something quite out of the normal.

By that I mean that our usual way of seeing things is violently reversed. When we stand in the valleys or lowlands we instinctively look straight ahead or up, and in doing so all opaque bodies are seen by their reliefs of shadows. We see these shadows and gain ideas of form from them, the eye finding rest in their dark depths by contrast with the occasional sharp breaks of high light. Looking down from a height involves a wholesale destruction of shadows, for we do not see them at all, and there is a consequent distortion of form. Every object is seen in its high light; not one is seen in its shadowed portion. More than that, the look downward means a monotony of light and a monotony of color. The direct sunlight is over all and is reflected back to us from every surface. Local color is bleached and changed by this, just as the color of a mountain-lake is lost in sky reflection. Finally, when we add to these distortions of the usual appearance the gray and hazy effect produced by seeing the world through a dense stratum of blue air, we have, I think, sufficient reason for saying that the view from mountain-heights, looking down, is not by any means the best view.

And, strangely enough, people on mountain-heights are forever looking down. If they would only look *up* they might see two features that are the better for being seen from high ground. I mean the sky and the clouds. The whole firmament seems to expand; and the curve down and around the world, given by the perspective of the clouds, is most impressive in its sweep. And what intensity of color in the violet-blue! What wonderful luminosity in the small, white cumulus and the feathery cirrus clouds! But this sky view is the one that people seldom see. They climb for the scenic view, which means a search for familiar objects on the map below them. In fact, it is more curiosity than a sense of beauty that prompts the climbing; for the most perfect landscape is seen from level ground, with the great sky space overhead as a leading feature.

The mountains themselves are seen at their best looking up from the valley. The view expanding, peak on peak, until finally the topmost spine is reached, is more complete than when one stands on the top and looks over snow-fields, down gorges and glaciers into the valley. The very grandeur of mountains lies in their height, mass, strength, and sky lines,

The look upward.

The clouds and sky.

and none of these is seen so well from the peak as from the valley. And here comes in the normal truth of color and shadow. Looking up, we have great masses of shadow broken by large expanses of light. Every cliff, every scar, every stone reveal them; and as the snow is reached, blue patches of it in shadow are contrasted with great pink fields of it in sunlight. Color is everywhere. The wall of the chasm is dappled with a hundred hues, the forests of pine stand in masses of dark green, the grass strips show pale green flecked with yellow, the glacier ice is blue-green, the rocks are gray, sometimes the needles of the peaks are dashed with cream-yellow at sunrise, or turned to pinkish-rose at sunset, and back of it all is the blue sky for a ground. The mountain's grandeur of bulk and line, its beauty of color and light are practically destroyed for us when we are standing upon the peak. We have, in short, the wrong point of view.

The mountains from the valley.

Mountain colors.

While not so impressive, perhaps, in their sense of loftiness, the mountains and ridges, that are but a few thousand feet high, and have no snow belts, are often more beautiful to look upon than the Alps or the Andes. Their tops may be turreted with rocks here and there,

The lower ranges.

but the jagged, saw-like effect of the higher peaks is gone. The sharp diagonal and the perpendicular line give place to the horizontal and the rolling line. These mountains are worn smooth by the elements, all the surfaces are rounded, and timber and grass grow readily upon them. But the mountain silhouette is still apparent in clear-cut rim; and everywhere trailing along the sky the eye meets the sweeping lines of ridge and promontory, or the billowy roll of descending lines flowing down by terraces into the valleys. How very beautiful these are in their undulation, as they join ridge upon ridge in rhythmical sequence! They twine and intertwine, curve and intercurve, weave and interweave along the sky and through the valley, until the whole fabric of the hills seems like a precious decorative pattern of green and purple embroidered on a blue-gray ground. There are no lines in nature more beautiful (save only water lines) than those of the mountains and the hills—particularly the untimbered hills, for as we descend from lofty heights the forms grow more graceful and rhythmical at every step.

And you who have, perhaps, lived for years with these mountains visible from your win-

Sky lines.

dow, have you noticed how they vary with
the different lights and atmospheres? Have
you seen them at sunrise lying off in the west,
when the light is on them instead of behind
them, and each barren crag gleams like a star,
when the pine forest on the ridge is pale and
blue, and the network of interblended lines is
woven faint and fleecy against the dark ground
of the half-awakened sky? How cold and still
are the valleys in shadow, and how spectre-like
the mists floating hither and thither, knocking
themselves to pieces on the mountain-side, and
finally dying out like smoke against the clear
sky! Have you noticed them at noon, when
the sun in the zenith has bleached their forest-
greens to grays and blues, when the valleys
drowse in the blazing light and the sky lines are
vague almost to the point of obliteration?
What a thick veil of silver-blue air lies in the
valleys and along the ridges, blurring and obscuring everything with delicate fingers until the far-off peaks seem turning into clouds!
The mountains lie enchanted under the wand
of the sunlight like the princes in Elfland.
No sound, no wind, no motion; silent they
rest under the falling light, reflecting the sky
above them. Of course, you have seen these

Mountains at sunrise.

At noon.

mountains at sunset, for then the light is behind them, and they stand in dark relief against a sky brilliant in color. The strength of an outline lies in its revealing the bulk of the body it encloses, and how well the silhouette gives the feeling of the mountain-mass! The shadowed side turned toward us is a great belt of cold purple, extending along from valley to valley, creeping up toward the crests, and seeming more purple than usual, perhaps, for the complementary yellow light that is above it in the sky. At twilight this range of mountains seems the division line between the world of day and the world of night. Deep shadow is flooding in from the east, brilliant light is in the west, and between them runs the dark mountain-barrier. It will light up presently under the pale glow of the moon, and the pines on the ridges will wave ghost-like in the blue night air; but now how shadowy and cool the mountains lie, and what a vivid contrast to the glowing heat of the firmament over them! It is one of the contrasts we all love, and however little people may fancy nature, there are few who will not turn to see the splendor of the western sky flaming above the mountain-ramparts.

If we shift our point of view, we shall see

effects of equal splendor when the mountains are lying to the east and are taking color from the last rays of the sun. They are far more brilliant in hue than the western mountains, struck by the light of early morning. The warmth of color is greater, because the sunlight in the morning passes through cool and clarified air, and at evening the same sunlight throws its light eastward through a heated and dust-laden air. The difference in atmosphere makes the difference in color and light, and these in turn make a decided change in the east-lying mountains at sunset. Indeed, form as shown in the outline and the shadowy silhouette is not now conspicuous. Lines are dissipated and surfaces are flattened into tints. The range may be shadowed at its base — a deep, hazy shadow —while the tops may be in full sunlight and receive the glow of the western sky on every bush and tree and crag with startling effect. The total result of reflected light from the range may be copper-color, pale yellow, rosy-red, or silver-gray ; and upon such a feature as a tall spur or bare peak the color may change from yellow to pink, from pink to gray, from gray to purple, until the light goes out of the west and the spur darkens and looms against the eastern

Looking eastward at sunset.

Mountain-glow at sunset.

sky like a cathedral tower, its edges lined with bright silver from the light of the rising moon behind it.

Beautiful as the mountains are under sunlight and moonlight, they are often more impressive under clouds with storm. The lofty majesty of the greater Alps in furious weather, their calm repose among all the turmoil of the elements, the mighty lift of the white, sunlit peaks out of the gloom of the valley, are sights never to be forgotten. The noisy winds, the sharp lightning, the torrents of rain that dash against the granite walls or hide from view the timber-robed sides, are mere sound and fury signifying nothing. The mountain is not moved. Secure in its strength, it holds its head above the storm and lives in the sunlight. Very beautiful indeed, are the tempests in the Alpine valleys, with darkness below and light above, and all that that implies of color-contrast. Nor does the scene suffer any by being mirrored in the greenish-blue waters of the Swiss lakes. Oftentimes the whole panorama of the upper air—cloud, lightning, green forest, gray rock, and sunlit snow-cap—may be seen in the lake darkened and deepened in tone by the local color of the water. The Alpine valleys may not be the spots of the earth that one would choose for permanent residence, because the sights they offer are too stupendous for daily contemplation ; but surely they offer the sublimities—the grander beauties of the earth and the elements—better than almost any other mountainous region.

A mountain is a mountain, and belongs to an order as human beings to a race, but there is quite as much of peculiarity in the separate peaks as there is individuality in different men. It does not appear so at first. We think all mountains are substantially alike, just as we think all Mongolians have the same features ; but a little study shows that there are never any two of them of the same form, color, or characteristics. Even ranges differ greatly in appearance. The Alps are not like the Andes, the Alleghanies are not like the Rockies ; and how different are the Harz Mountains, with their green slopes and cold blue air, when compared with the bare Tuscan mountains, so positive in their light and warmth ! Wherein lies the individuality of the isolated mountain it is difficult to say. It may have height and arrowy dignity compared with its squat, smooth, or ragged neighbors ; it may be distin-

Mountain individuality.

guished by forms of timber, rock, or grass, but these features often undergo odd changes with various lights. The mountain lines against the sky change also as we change our position. We think we know the profile until we see it from a different side and in a different light. The Man's Head, the Anthony's Nose, or the Devil's Pipe, outlined by some projecting crag against the sky, has nothing to do with mountain individuality, though it may have to do with local name and identity. Such fancied marks lose all likeness as soon as we move away from a certain position. Even the little hills have a way of tricking us with different aspects; and every hunter in the Bad Lands who has made a mental "blaze" of a butte on his trail knows how often he has failed to recognize that butte when coming upon it from a new direction.

Changes of form.

Bulk and mass also have some influence in marking the mountain, though these, too, apparently change as we shift our standing ground; and color gives some distinct character, yet this is, perhaps, the most inconstant of all mountain features. There are few things in nature that can show distorted color so well as a mountain-top under sunlight. The light is continually bleaching or heightening

Changes of color.

the color in such a way that it appears odd to our eyes. This is peculiarly true of the noonday light, which flattens a dark stone-color to a silver-gray, and will turn a belt of pine timber from a dark green to a pale blue. Finally, there is always some difference in mountain appearance, dependent on the thickness or thinness of the atmosphere, to which must be added allowance for the distortion caused by the top of the mountain being usually observed through a thinner layer of air than the base. *Influence of atmosphere.*

When all these features are considered, the mountain instead of being a steadfast, unvarying tower of rock is, to all appearance, one of nature's fickle creations. It shifts countenance as many times in a day as the sky above it. One moment it is blue under direct light, the next it is green under cloud-light; at dawn it is gray; at sunset it may be golden or even red; at night it is cold purple. The changes are less marked on a cloudy day, and a mountain's bulk, height, and surface are seen to better advantage then—yes, even on a rainy day, when clouds are hanging about the peaks—than under sunlight. *Light changes.*

A hill, as we have already noted, may be a

The green hills.

worn-down mountain, or it may be mere glacier push, or again, it may be a hard core of rock that has defied the wear of water. The variety of hills is even greater than that of mountains. Often they are banded together in groups or chains and dignified with the name of "mountains;" sometimes they are in clusters and lie nestled together along a river's course; and sometimes they rise singly from a flat basin or plain. They almost always show the effects of erosion, and, indeed, the marks of the streams about their bases and sides can be easily traced. The tops and sides, washed by rains, have enough soil for vegetation, and trees or coverings like the heather grow readily upon them. Every country has its different kinds of hills, and in Great Britain almost every shire will show a new species. The bare cliff-hills along the English Channel near the Isle of Wight, so clear and pure and beautiful in their sky lines, are different from the rugged hills of Scotland, with great bowlders sunk in the purple heather of the peat-beds; and every traveller must have noticed the change from the flat hills of Suffolk to the abrupt ranges of Derbyshire.

English hills.

The damp climate and the heavy rainfalls of

Great Britain have made it a country of hills. Nowhere are they to be seen in such beautiful combinations, and to them England, in particular, is much indebted for its beautiful scenery. In all seasons, in foliage or with snow, the country of low hills is an attractive country. The charm of Normandy and the Rhine provinces, as of New England, lies in the broken, undulating surface. To whatever point of the compass we turn there is unity in variety. The amphitheatre of hills surrounding Amherst in Massachusetts does not grow monotonous to those who look out upon it from day to day. The encircling parapets always have a new tale to tell, a new wonder to reveal. No sun gilds them twice in just the same way, no atmosphere is repeated for any two days, and the mantle of green in summer, the robe of white in winter, are never the same from year to year. *New England ranges.*

Fortunately enough the round-topped hill, upon which the Assyrian built a city and the mediæval baron a castle, is to-day left to nature to do with as it pleases. The modern builder places his city on flat ground, and if there is any little mound in his way he levels it into any little valley that may be near at hand. He wishes everything flat and squared in his city, *Hills and civilisation.*

though from his own door-yard he likes well enough to see the hills in the distance. And in the distance they lie covered with grass and timber, gladdening the eyes that look at them. The cattle go to them in the heated season, as the birds in times of cold and storm, and down their sides of moss and rock run the little streams that keep the valley green and turn the mill-wheels of the factories. They are always beautiful, breaking as they do the horizon line with new forms, new colors, and new lights. And we need not be disquieted about them because they are worn-out mountains and must eventually become flat meadows. True enough, they are passing away. The bare butte of Montana is slowly sinking into a lump of formless clay because it has no covering to shield it from the elements. The New England hills and the hills of Old England are sinking, too. It is nature's plan to beat down the mountain into the dust of the plain and the sand of the sea-shore; but the plan will take many ages for its fulfilment. To-day the little hills clap their hands and rejoice as in the days of David. They will not disappear until another David comes.

The levelling-down.

CHAPTER XII

VALLEYS, PLAINS, AND LOWLANDS

THE lines that give character to the mountains, the valleys, and the plains also create in us definite feelings or impressions. When nature shows us the broken or abrupt line we gain from it an impression of activity or restlessness; when we see the long, diagonal line the impression received is one of swift movement, as in the downward flight of an eagle; when the flat, horizontal line appears the impression is one of rest, peace, inaction, even drowsiness and sleep. Hence it is that people speak of the abrupt and broken mountains as representing the earth's action, while the low-lying marshes and meadows represent its repose. There is a truth of feeling or imagination in this. The broken peaks and spurs, jutting up from the mountain's ridge against the sky, certainly do seem restless, suggestive of motion; while the meadows, where flowers grow and bees hum and cattle recline at noontide under the trees, are

Line-impressions

just as truly suggestive of listlessness, idleness, and sleep.

These impressions produced by nature's lines are doubtless wholly subjective, yet they seem positive realities to us. A man can no more rest on a mountain-peak than he can sleep standing upright. The perpendicular affects him one way, the horizontal quite another way; and rhetoric has not erred in speaking of the "restless" mountains, though they are as motionless as the plains; nor of the "sleeping" valley, though a valley never sleeps or wakes. Perhaps the chief characteristic of the valley is its repose. It is always still, except when set whispering with winds or roaring with storms; and the deeper, the more shut in it is, the greater seems its hush. Standing above it at mid-day, with light and shadow lying along its sides, the stillness seems like the silence of untenanted space. A rifle-shot or a human voice breaks upon the sensitive air with a sharp crash, and the echoes set flying by it reverberate and pass out of the cañon ricocheting from rock to rock with the elasticity of a rubber ball. Quite a different affair, too, is the sound of thunder in a mountain-valley compared with the thunder heard on the plains. The clap and

The "sleeping" valley.

Valley silence.

peal are terrific; the roll from side to side is repeated again and again, until at last it dies off up the gulch in a muttering rumble that shakes the whole atmospheric envelope. It is only an accidental affair, and as soon as the storm has passed, the valley once more addresses itself to sleep. The mountain-shadows lie clear and cool along the ascending slopes, and as the valley drowses the day through, these shadows grow longer, each one stealing silently down the western side, crossing the valley-brook, and creeping up the far eastern slopes as the sun sinks down beyond the mountain-peaks.

Echoes.

And what masses of shadow there are in a valley! However it may lie as regards the points of the compass it is always sure to have its slopes, its hills, and its mounds that cut off the sun's rays and create the dark-green patch. Even where the valley is quite wide, the timber that usually grows thick in the basin creates its own shadow in an almost impenetrable screen of foliage that shuts out the sun. These forest shadows are usually dark, moisture-laden masses, deep green in hue, and seldom marked by brilliant colors. In fact, the mountain-valley is not the place where nature

Shadows in the valley.

puts on her most gorgeous garments. Occasionally, in what are called the "sunset" valleys—that is, valleys running east and west—there is some warmth of color, and in autumn, with the yellow foliage and the Indian-summer haze, there is often great display; but during the hot months the predominant note is green, save where in the distant gulches and *coolees* the blues and purples assert themselves.

There are two ways by which the mountain-valley may come into existence. The first is by the cutting-away process of torrents; the second is by depression. Oftentimes the heave of the fold that has lifted the mountain-peak skyward has allowed the valley to sink back and downward. A depression is thus formed which the wear of water immediately increases. Some valleys are even sunk lower than the surrounding country—so low as to make a hollow—and in time the waters flowing into them form the long, twisting mountain-lake of which the Lake of Lucerne in Switzerland is an illustration. More often, however, the valley is elevated above its surrounding plains. From its walls one can always gain an approximate idea of its age, as from the peaks one knows the age of the mountains. Abrupt sides

"Sunset" valleys.

Valley formation.

Age of the valley.

of rock, with precipices and overhanging crags, prove one of two things : Either the rock is very hard or the exposure is very new. The wear of the elements tends to round, smooth, and flatten down all such sharp projections. In the older valleys of the world, such as those of the Alleghanies, the sides are sloping, the basins rounded, and the lines against the sky show only the smoothest curves. Usually a small river or brook winds its way down the larger valleys, cutting out the soft deposits of earth and forming banks or cliffs on either side, where vines clamber and stunted pines cling in the fissures of the rocks, and small trickling streams drip from under thick carpetings of moss. It is usually a noisy, swift-running stream, dashing its way seaward over shelves of stone and gravel, winding in and out of deep pools, and swirling around sharp bends in eddies and circles. Its tributaries are the little cold-water rivulets that come down the side gulches, springing over ledges and bubbling into basins—streams where the young trout splash in their leaps up the falls, and where the stealthy-footed inhabitants of the wood come to drink.

Sloping sides and smooth curves.

The brook again.

The brook, the river, the valley, and the

mountain-walls all have their special features that attract; the brook in its flashing motion and light, the valley in its mass of foliage, the mountain-walls in their color, their shadows, their bulk, and their lift against the sky. All of them are seen at their best during the months of summer. In October, when the autumn leaf is rustling, and the rain begins to fall on bare boughs, a strange feeling comes over one in looking at the valley—a feeling that its bright days are numbered, and that it will soon be sleeping under ice and snow, with its protecting mountains looming dark and grim through the long nights of winter. But at any time of the year, and with all the beauties the valley may reveal, it is not the best place for habitation. The conditions of life are harder there than on the flat-lands; and the density of the shade, the jungle quality of the foliage, the enclosing walls of the mountains, are all oppressive—in a way stifling and stunting in their effect. The very animals and the birds seem to feel this, for they are not so frequently found here as upon the edges of the flat plains, where the country is open. Man himself grows rather heavy and stolid when hemmed in by mountains, or sur-

In autumn

And in winter.

rounded by heavy timber. The open country, where the sun shines through the shade, where the soil is free from rock and the tree from moss, is the better abiding-place. In such a country man moves hither and thither with greater ease, the climatic conditions are more endurable, the earth is more arable, the rainfall more equable.

The valley home.

As we descend from the mountains this open country first appears in the table-lands or upland plateaus. They are usually high above sea-level, sometimes several thousand feet; and in appearance they have something of the rugged-broken surface characteristic of mountains, mingled with features peculiar to the prairies. These table-lands are often open, treeless regions, and are generally arid. The atmosphere above them is dry, and so clear that objects appear nearer than they really are—outlines of mountains, for instance, showing very distinct, though many miles away. On almost all the high plateaus distances are deceptive, lights are brilliant, and the blue sky above glows with a wonderful intensity, and not infrequently with a violet tinge about it.

The table-lands.

The Montana table-lands are perhaps exceptional. They are full of abrupt breaks, with here and there sawed-off mountains that are succeeded by flat basins, where once the buffalo grazed in countless numbers, and where even to-day one may occasionally see the sheeny coat of an antelope glistening in the sun. The eastern portion of the state bordering on Dakota shows in its cliffs, buttes, and gravel beds a land once shaken by volcanic convulsion, and water-swept by flood and glacier. Timber is rarely seen upon it, grass grows in small tufts but a few inches high, and the predominant growth is sage-bush and cactus. Yet its weirdness and its desolation make it attractive; and to one interested in color it is the queerest region in all the world. The dry, alkaline clay throws off local hues of red, orange, pink, and yellow with the first glint of sunshine; and the shadows are blue, violet, and lilac. These are the same hues of decay that we met with in Venice, for the Bad Lands region died centuries ago. It is to-day showing us that beauty of color which we see in iridescent glass, and the cause of the one is the cause of the other; that is to say, the disintegration of fibre, the chemical rot of matter.

But the Bad Lands country is something of an accident of nature—a tumbled and broken

district isolated from the table-land family.
The Arizona and the Colorado countries are
very different from it, and neither of these
bears much likeness to the Asiatic table-
lands, like the steppes of Siberia or the great
plateau of Tibet. All of them are fine in
horizon and mountain lines, in skies, and in atmospheres. All of them again have picturesque
spots, where swales and basins fall into graceful
shapes, where water runs, and grass grows. And
again, all of them are stimulating in their wildness and aloofness from civilization. These are
the primeval tracts, never subjugated by the
plough—the free spaces of the world, where the
wind blows up and over the hills and ridges,
blowing toward No Man's Land. The feeling
of solitude, of being alone with nature, is omnipresent; and there is enough of the savage in
everyone to feel pleasure in that sensation. We
may aspire to the stars mentally and spiritually,
but nature made our feet to tread the earth.
The animal in us cannot be wholly eradicated
by any course of ascetic training. I have seen
wild horses on a high ridge snorting with de-
light at the sun and the wind; given the opportunity, the physical in man will assert itself
just as strongly.

Plateaus and steppes.

The primeval tracts.

The prairie.

Lower down than the table-land comes the American "prairie." It is not so abrupt in form as the table-land, but was once very like it in the feeling of wildness which it fostered. The name was originally given by the French *voyageurs* to the flat plains of Illinois and Indiana; but it has been applied of late years chiefly to the long rolls of land that stretch across Dakota, Iowa, Nebraska, and Kansas. Their forms have been likened to the great swells of a tropical sea. The land rises in crests called "divides," and sinks into hollows called "swales." The grass once grew rank and tall on these prairies, bending and rolling in the wind, but from their earliest discovery trees have been known upon them only in isolated spots along river bottoms.

Treeless tracts.

The absence of trees on these fertile lands has never been satisfactorily explained. They grow there readily enough to-day when planted by man, but for centuries nature planted and grew nothing but grass. It is said that the burning of the grass by the Indians, to drive game, destroyed the timber-growth, but the explanation is of doubtful value. The great conflagration of the plains that the Indian novelist has told us about, is at best a lively piece of the imagination. In certain areas there have been fierce blazes with high winds; but the hundred-mile sheet of flame that travelled faster than a horse could run and led up to the dramatic race for life, is something that no one—not even Kit Carson —ever saw.

Prairie fires.

The continuous rise and fall of the prairie divides and swales, as one rode over them years ago, could hardly be called inspiring. To see the sun come up from the grass and go down at night into the grass again; to see one's horse walking shoulder-deep in it, and to watch it bending before a fast-travelling gust of wind, its surface changing in greens and yellows like a changeable silk, were novel sights at first; but they finally became a little wearisome. The lack of shade, of hills, of valleys, of trees, of water, was keenly felt. When chance brought one upon a prairie pond fringed with tall rice, where wild fowl were flying hither and thither, the change was almost like coming upon an oasis in the desert. Even the round dry basins of the prairie where in the old days the buffalo made the night circle against the wolves, or the deep trench caused by cloud-bursts, proved of exceptional attractiveness after miles of travel through that rank-growing grass.

The roll of the prairie.

But the prairies have undergone great change, like all things American. The settler and the plough have turned under the Indian and the buffalo, the divides are now planted with houses and wire fences, and the wind is blowing over fields of wheat instead of prairie grass. The great charm of the land, its wildness, has passed away. Time was—and not more than thirty or forty years ago at that—when never a trace of white man's activity was seen on the Dakota uplands; when not a railroad crossed it, and even an Indian trail was almost unknown. The horseman found his way by the run of the divides or the sun, and every adventuresome explorer riding over that tract felt in his heart that he was another Balboa discovering the Pacific of the plains.

The prairie wildness.

It is not impossible that that wildness may return again, for nature has a way of reasserting herself after long bending to the will of man.

Nature's revenges.

> " They say the lion and the lizard keep
> The courts where Jamschyd gloried and drank deep;
> And Bahrám—that great hunter—the wild ass
> Stamps o'er his head, but cannot break his sleep."

Those who have been plucking the brightest skeins from the fabric of the prairies will pass away, and when their fingers are stilled the great Penelope will once more speed the shuttle. The prairie grass may wave again when the ploughshare is beaten to dust and the Dakota village, even in its ruins, shall have perished. Nature will come to its own again, for during all these centuries of man's dominion on the earth it has not ceased to whisper in the ear of history: "They shall build, but I will throw down." In its own good time, the ravaged prairie will be re-covered with a mantle of waving green; the by-ways and the haunts of man will be obliterated, and the sun will shine, the wind will blow up and over the divides and swales, blowing once more toward No Man's Land. *The wilderness again.*

The flattest plains in the world are those that have been at one time the beds of vast inland seas or lakes. The plains of Hungary are of this type. The largest one is now drained by the Danube, and is not remarkable except for its marshes, through which the river winds. It is not very different in appearance from the ordinary coast-lying plain, which is to be found in almost every sea-bordered country. Properly speaking, the coastal plain is a tract of land reclaimed from the sea, either by the slow up- *Flat plains.*

heaval of a low-lying shore or by the gain of silt washed down to the shore by the rivers—something won from the sea either by upheaval or accretion. Holland is an exceptional illustration of a marine plain reclaimed from the sea by human ingenuity aiding the favorable drift of sand into dunes along the coast; the State of New Jersey, or at least a part of it, is an illustration of a gradual upheaval that has placed the plain above sea-level. I do not know the geological formation of the east coast of England, but I suppose it to be a plain similar in origin to that of New Jersey. These tracts now lie above inundation, and are broken by low hills, stretches of meadow and timber, and slow-winding streams. They make the arable and the livable portions of the globe, and in many respects they are the most picturesque portions. The flat horizon lines, the great sky depths, the feeling of space, the expanse of light and color in the sky, are all features that are not impressive at first, but soon become attractive and finally most lovable.

The lands subject to flooding by high tides (perhaps the coastal plains of the future, now in process of formation), called marshes and meadows, are common enough along every coast

Livable lands.

The marshes.

where rivers empty into the sea and silt is washed down. The Atlantic coast of America, from Massachusetts to Florida, has a plenty of them. They are almost useless for human occupation, and though the soil grows a rank vegetation, it is not edible for man or beast. Because they cannot be utilized to advantage, they have been regarded with some contempt by mankind; and the preacher, the orator, and the poet have always paralleled them with human stagnation or vileness. But they do not deserve such odious comparisons. Humble and peaceful under the falling sunlight, they have their share of the universal glory, and were constructed by nature for a useful purpose. They are the outer fortifications of the coast, keeping back the sea, and growing strong vegetation to prevent the wear of water on the land. How unsightly would be those lands if it were not for their thick coverings of reeds and rushes! How beautiful are they now garmented in the pale golden-greens of spring, the emerald-greens of summer, or the golds and browns of autumn! I have seen ordinary marsh flags with a low, summer sun behind them, when every blade looked as transparent as cathedral glass, and every leaf-edge was showing the colors of the

How characterised.

Reeds and rushes.

spectrum. And again, under the morning sun, with the wind blowing over them, I have seen them glitter and throw light from their polished surfaces like the bayonets of a regiment on parade. And still again in mid-winter I have seen these same commonplace flags standing yellow as gold above the snows, with every stem casting a bright blue shadow, and the whole scene of marsh, sky, and snow showing a perfect color-harmony in yellow, blue, and white.

Indeed, there are many beauties that adorn these marshes unseen by the man who wades across them shod with rubber boots and carrying a gun in his hand. There is something quite as beautiful as wild fowl to be seen from the sunken "blind" on the point of land. The play of light on the flat mud near the water, the scarlet sky reflection on the little waves, the amethystine hue made by a flaw of wind rippling the surface of the bay, the splendor of the sky, the radiance of the white clouds, are all incomparably fine. Looking backward, the rushes of the marsh extend for miles in one great sweep of color, till they meet the woods, and beyond and above the dark woodland mass stretches another sweep of deep blue sky. There never was a simpler or a nobler landscape.

These marshes, whether seen in the summer, when they are so luxuriant in their greens, with the flag in blossom and the young cat-tails nodding in the breeze, or in the fall, when nature is dying and the reeds are day by day shifting through green to gold, when the trees are gorgeous with autumn tints and the orange stain of the short grass is gathering and growing and weaving itself into a brilliant carpet whose colors do not fade until after snow falls— seen, indeed, at any time of the year, they are far from being the pestilent congregation of vapors and malaria which fancy usually pictures them. Even those marshes that lie close to cities and have ramshackle factories scattered over them, like the Hackensack meadows— marshes that are damp with mists and fogs and thick with smoke and dust—even these have their charm of color, broken light, and atmosphere. In picturesque qualities they are almost as fine as the dunes and meadows of Holland. *Color changes.* *Near to civilization.*

In abbreviated proportions the same lowlands line the shores of almost every large river, particularly the rivers with broad basins. The rushes, reeds, and wild rice grow there even better than by the sea. Along the Mississippi the low, flat spaces on either side of the river,

The "bottom-lands."

called the "bottom-lands," are taken up by marshes, timber-growth, and lakes. Sometimes the lakes with flags surrounding them look like the shore regions near Chesapeake Bay; but more often the bottom is a vast jungle of trees, vines, and dense undergrowth, not unlike the Dismal Swamp of Virginia. Its impressive feature is its luxuriance of vegetation. Its trees are often enormous in size, the grass stands higher than one's head, and the ground is black with the mould of centuries.

Swamp-lands.

The sloughs, or little water-ways connecting the lakes or marshes, run sluggishly in blue-brown streams, and the density of the shade scarcely allows of much sky reflection in their coloring. Sometimes an open spot in the timber, where wild rice surrounds a small, shallow lake, gives a bright dash of sunshine and color; but as a general thing the bottoms are not brilliant in hue or attractive in light.

CHAPTER XIII

LEAF AND BRANCH

It is commonly stated in the encyclopedias, I believe, that the lakes of North America contain half the fresh water on the face of the globe, that the rivers of the Western continent are the largest and the longest in existence, and that the whole area of North and South America is the best-watered and the most fertile land in the world. The truth of this statement granted, it should follow that the land of the two vast countries is more productive of vegetation than any other known to man.

The New-World vegetation.

This is not merely an inference, it is a statement of fact. The palms of South America have a maximum height of from one hundred and fifty to two hundred feet, the red-woods of California are sometimes ninety feet in girth, and how tall were once the pines of the Northwest woods I cannot now say, but their ranks were countless, and they covered millions of acres. It is true that these are growths of ex-

The foliage.

ceptional size, things of rarity; but if we consider the density of the ordinary woods, the thickness of the undergrowth in every commonplace valley and on every hill and mountain-side, the leafiness of the foliage in this Western world becomes almost appalling. No dweller in the Eastern United States, who is content with a vacation in the Catskills, the Adirondacks, or the White Mountains, can have more than a faint idea of it. He is looking at second-growth timber honeycombed by the axe, at fields broken by the plough, at hill-side thickets eaten by fire. The sparse remains of the primeval forest in the Northwest, the timbered valleys of California and Oregon, the vast woods of Alaska, tell the tale of what America once was, and would be yet, were nature allowed to build undisturbed and as it pleased.

Timber-growths.

All members of a series, yet how varied, are the families of trees, and what a different landscape effect they produce when massed in groves or woods! Almost every valley, hill, and upland in America presents an appearance peculiar to itself by virtue of its timber-growth. The giant red-woods of California, the great elms of the Mississippi, the cotton-woods of the Missouri, the oak-openings of Minnesota, the Eastern tangle of wild cherry, hickory, and beech, have little resemblance one to another. And those long aisles and open spaces in the forest of oak and chestnut—spaces where the sunlight breaks through in splashes, where the creeper grows and the cardinal flower gleams— what a contrast they are to the dark depths of the "pinery," where the closed-up ranks of the trees shut out the light of the sun, where the long moss hangs in festoons from the branches, and only stray patches of the lowly pink peer through the carpet of pine-needles! *Variety of forests.*

But deep forest and dark pinery are hardly attractive to the average person. People have some fear of the shadow and the solitude, and quickly wish themselves back in the sunshine with friends. They prefer the more open groves, where the light breaks in flickering bars across the wood-road, where the field of golden-rod is in sight, and the blue sky is not shut out. Certainly the open woods are the most enjoyable, the most livable spots; yet those great interlaced forests where light filters through only in arrowy shafts, where the bear and the wolf slink like spectres and the deer breaks suddenly from his bed—those labyrinths through which stretches no Dædalian thread *The depths of the forest.*

—how sublime they are in their power and volume! To the uninitiated and the timid they may have terrors, but to the hunter and the backwoodsman the "big timber" is an earthly paradise. There nature is supreme and man is only a cipher; there heat, light, and moisture work their pleasure undisturbed. Within the pale of civilization, upon meadow, field, and hill-side, one can never feel that nature has done justice to itself or its growths. The woods upon our Eastern hills have all been raised upon the bottle. Where the Great Mother is unthwarted in her ways, she rears a brood of giants.

Big woods.

The botanist has classed, ordered, sectioned, and specied the different trees, and christened each with a Latinized name; but I have no thought of following his scientific arrangement nor of cataloguing or classifying the different varieties of trees. My task has to do with surface appearances. Moreover, the general character of a tree is revealed by its form, color, or texture; and it may be assumed that the average person recognizes it by these features rather than by reducing it to botanical class and species. How much depends upon outline, hue, and surface, and what distin-

Botanical classes of trees.

guishing ear-marks these are, may be suggested by a few haphazard descriptions of the common trees about us.

The spruce, for instance, is a straight-trunked tree that throws out branches that ride upward like crescents, and bear needles that hang downward like fringes. Its outline, when seen in silhouette against the sky, is pyramidal; its color is dark green, often blue-green when seen from a distance, and at twilight it is cold-purple. The pine is like it, but its branches are not so crescent-shaped, and the needles push outward in clusters rather than droop downward in fringes. It is of a darker color than the spruce, and at night or under shadow it is bluer. The poplar is a tall tree, and often a straight one, but the branches do not swing outward like the pine. They seek rather to grow straight beside the parent stem, and the twigs and the sharp-pointed foliage surround the branches as a loose sleeve the arm of a woman. It is white-trunked, with a leaf that is bright green on one side and silvery green on the other side. The black oak grows a straight trunk with limbs that shoot out almost at right angles; but the white oak and the pin oak are crooked and twisted, their

Tree characteristics.

Various tree-forms.

harsh trunks are often broken with boles, and their limbs may take angle lines or prong out like the horns of a deer. Very different from such an angular growth as the oak is the stately elm, its long limbs branching and falling so gracefully, the weeping willow that throws its branches up and over like the spray from a fountain, the round, ball-shaped horse-chestnut, or the long-armed, white-breasted birch of the mountains.

Branch ramifications.

The locust, the sycamore, the tulip, the linden, the nut-trees and the fruit-trees are just as individual and peculiar in their forms. The most commonplace hill-side will show innumerable classes, families, and groups of trees; and to the romanticist many of these growths convey significant meanings by their forms or movements. It is doubtless an application of the pathetic fallacy to think of the willow as "sad," and yet the droop of its branches, the wave of its leaves, lead the poets to make such a statement. In the same associative way, the pine on the mountain-top is "solemn" or "lonely," the yew and the cypress are "mourners o'er the dead," the oak is the "monarch of the forest." Their look and bearing suggest such descriptions; and it is not strange that man should

The pathetic fallacy.

sometimes assign to them attributes peculiar to humanity. A century-old oak has about it something more than sturdiness and bulk, it seems to have dignity, nobility, and fortitude. How proudly it stands against the elements, and how nobly it was designed to stand! Its roots are driven deep into the rock ledges; its massive trunk and branches are constructed to endure all weather. It has sensation, and it seems almost human as it stands there year after year, changing its garmenting with the seasons, sighing as the wind passes through its branches. And how serenely it lives and dies! The growths of nature are in no way hurried. Time is a human check-system of which the bud, the leaf, and the branch know nothing. They grow to maturity, and pass on into old age and decay with patience. The oak has its portion of earth-glory, something of beauty in light and color, something of usefulness as shadow and screen. These it receives, reflects, reveals; and having fulfilled its destined end, it sinks back to the earth whence it sprung, never questioning the reason of life or the wisdom of death. Such personification is no doubt mere pathos and fallacy; and yet, for all that, there seems to be a nobility

The so-called sentiment of trees.

Life of the oak.

about the oak. At least such is the romanticist's point of view.

The only power of motion possessed by a tree lies in its growth upward, downward, and outward. It is capable of being moved, however, and the great mover is the wind. The slender trees like the birch, the willow, the elm, and the maple, are swung and tossed in their branches as well as in the upper parts of the trunk; whereas the sturdier growths, like the oak and the chestnut, are moved only in their leaves or smaller stems. In a heavy gale the large trees often rock when they will not bend. The pines, the spruces, the hemlocks—all the conifers—are great rockers. And they are also great whisperers, great musicians. The slightest wind will start the white pine sounding its Æolian harp of needles, and in a gale the whole tree will sometimes hum like the wires strung on telegraph poles or the wind-swept cordage of a ship's rigging. The elm is one of the most graceful of the bending trees, and in fresh winds its branches will roll on for hours, an epitome of poetic motion. The birch is still more easily bent, and the very word "willowy" indicates the elasticity of our common meadow-tree. The poplar, though often a tall tree, is somewhat stiff in its branches, but it hardly knows such a thing as rest in its leaves. The slightest breeze starts them trembling. The Normandy poplars are forever fluttering and twittering, even in calm weather. The gentlest breath of wind will turn up the silver of their foliage, and a row of them along a road will glitter and flash light at times like the glass pendants of a chandelier. Strange flashings of light and color are also shown at times by the beech, particularly the copper-beech, and it, too, sways easily ; but not so the large-leafed trees like the walnut and the oak. They make much noise, but move less in their branches than the thin, narrow-leafed growths.

Leaves in motion.

The leafy trees in groves or forests, when agitated by winds, have a sound like that of a distant waterfall or fast-driven rain, and anyone who has stood on a mountain-top and heard a storm coming down the valley knows wherein "the roar of the storm" consists. It is the roar of foliage struck by wind and rain. All the sounds from trees seem to be more subdued at night than at any other time. The night winds that stir the leaves and set the whole wood whispering, are gentle breezes, and possibly because of their gentleness they are great

Roar of storm.

Winds in the forest.

creators of sentiment. The sound is not only restful, but under moonlight when the dark shadows of the wood seem doubly mysterious, it is suggestive of music, poetry, memories, love, life, death—all things of passion and of beauty tinged with sadness.

Bare boughs.

It is a great change from the summer breeze and the barred moonlight on the wood-path to the windy days of March, when the bare branches moan under storm-skies and the sere leaves of the oak grate dryly on their brittle stems. It is not the season of poetic sounds, but it is the best time of all the year to study the trunks, boughs, and branches of the trees. Indeed, the windy March has always been reviled in the name of the leafy June; and yet it is a most interesting month, full of promise, full of graceful lines, full of silver-gray beauty. The trees stand stripped and bare, the trunks are blackened and weather-stained by winter rains, the twigs have not yet begun to redden under the ascending sap; but how beautifully the branches ramify and spread; how tenderly the little stems bunch up together or are etched in dark lines against the sky! What contours, what delicate light-and-shade, what infinite grace of line these bare branches show us!

And in March how strong the bare forest breaks across the horizon, how clear and sharp the dark ranks along the hill-top cut the sky! The iron-like trunks show a variety of darks, though to the casual observer they are all of one tone; the twigs that bunch together broom-like along the top seem like a bordering fringe; and the dull-green mass of the cedar,

In March.

"That keeps his leaves in spite of any storm,"

is merely a color-spot in the line. And how all this outlining of the woods in detail and in mass fits in and holds its place in the envelope of the landscape! Nature is stripped of its gay garments. It is showing more of structure than of color. The lines of shore and hill and mountain, of tree and field and rock, are everywhere apparent. A cold light cements them all, and it is the *ensemble*—the unity of many in one—that makes such individual parts as the bare boughs and branches appropriate and beautiful. We may prefer certain months, lights, skies, hues; but the cold sky and light of March belong with the leafless earth and harmonize with it just as completely as the red foliage of September with the yellow-flushed sky of Indian summer.

The March harmony.

Warming color.

In a few weeks there is a very noticeable change in the whilom March woodlands. The young trees begin to show dull red in their smaller twigs. A reddish hue spreads all along the bordering fringe at the top. It is the first positive color-note of spring, though in the small trees and bushes it is seen all winter long. As the warm sun starts the sap the color begins to brighten. The swamp trees with their roots in the water show it first of all, and then others join in until at last there is a distinct hue of dull red running through all the woods. The buds swell and begin to open just a little, a fuzziness muffles the sharp outlines of the branches, and the next color-note is a mist of pale yellow, mingled with the pinks, grays, and whites of the buds and the reds and yellows of the stems. A few weeks more and the leaves are out enough to thicken the view, obscure the tree lines, and cast a yellow-green hue over the forest.

The budding season.

The grass has at this time grown long on the meadows and is deep green in color, but the foliage of the trees comes later. The chlorophyll in the leaf-cells is not strong enough yet to show the dark green of midsummer. The young leaves are all tender in hue, shiny, coated with a varnish at times; and many are the transitions through which they pass before they gain their summer coats. The maples and the willows are the earliest ones to leaf out, and the oaks about the last ones. When the first leaves of the white oak shine against the blue sky like blossoms, almost all of the other trees are far out in foliage, and yet the buds of the black oak and the hickory are just beginning to break. *Summer foliage.*

Early in July the leaves have reached maturity in size and color. After that they change little for two months, except that some of them grow more shiny and others again appear to dim their lustre. The character of the tree as portrayed in texture and color is now well developed, and the delicate honey-locust, the leathery-leaved hickory, the drooping willow, the shaking aspen, and the copper-beech are all in their prime, contrasting with and relieving one another in the landscape.

The massed foliage when seen on cloudy days during the summer months is dense, dark, and bluish; on clear days it is bright green, and under strong sunlight, often fire-green. The predominant note everywhere is green, but it has its thousand varieties in tints and shades, and each one of these has a gamut of its own, *The variety of greens.*

Light transformations.

which it runs over daily with the shifting of the sun. Light transforms all things, and we have already seen what changes it may produce upon the foliage of a mountain-top. The leaves are heightened, deepened, bleached, or distorted, according to their texture or light-reflecting capacity. Often the green of a tree-top is turned to cold gray under a noonday sun, and at sunset, when the trunks of the trees are in shadow and their tops in full sunlight, everyone knows what a sharp contrast appears. The top is yellow, the body dark green. If the tree has a glossy leaf, the whole top may be a mass of reflected light. The tall tulip, the sycamore, or the chestnut at evening, with its loftiest leaves apparently changed into small shields of flashing light, is not an uncommon spectacle.

Swift color changes.

I fear that many of us have small conception of the changes that may take place in a green leaf in the course of a single day. It is green in our hand, and we naturally think it must be green on the tree; and so the easy conclusion is reached that leaves in summer are green and never anything else. But they are seldom the same green for any length of time. I once tried to keep a record from day to day of the color-changes in a few lawn trees, but the attempt had to be abandoned. The shiftings of color were so frequent, owing to the changes of light, that the notes were apparently conflicting and led to no result but contradiction. Even with the so-called "flower-bearing trees" like the tulip, the locust, and the orchard fruit-trees, the color-transitions from hour to hour are swift. Beautiful, indeed, are the white blossoms of the cherry, the pink-and-white of the apple, the darker pink of the peach. Seen in the early part of May, before the foliage has opened, they make charming masses of color along the hill-side of the farm and against the woods. They are tokens of the winter passed and the spring arrived, and while they are swaying cloud-like in the orchard, the castellated cumulus is piling higher and higher in the glowing sky. Fair things of spring, beautiful they are while they remain with us, but how quickly they pass! The blossom of to-day is not the blossom of yesterday. Its color and light are different. And then some night the wind rises, a "blossom storm" comes on, and in the morning the light and color lie broken on the ploughed ground and the dark boughs look more desolate than ever.

Trees in blossom.

Blossom storms.

Autumn glory.

And what of the autumn glory of the trees! What of the changes here that mark the ebbing season, beginning with the first maple-leaf that turns yellow in September and ending only with the dark, wine-red leaf of the oak left fluttering alone against the blue sky of December! Out of the green of summer, into the yellow, the pink, and the red of autumn, the great procession moves. The chlorophyll has exhausted its power in the leaf-cells, the green is bleached, the yellow must follow, and finally the russet of decay. The transitions are even but rapid. The different stages come and go, the hues passing from one into the other so softly, so easily, that before we know it the whole face of nature is changed and the panorama of the scarlet fall is before us. How swiftly the days fly, and when there comes that lull called Indian summer, how we wish it would last forever! But the great globe spins like a potter's wheel, and the coloring that this week stains the valley of the Hudson with carmines and saffrons, will next week be shifted southward to the shores of the Delaware. The splendor moves with the sun, northward in the spring, southward in the autumn. A fortnight or more and the gorgeous leaves of the hills, torn by the

Indian summer.

storms, will be flying with the winds, heaping in fence-corners and about bushes for the long, long-sleep of decay; but while the flame-like mantle lasts how supremely glorious its coloring!

The distribution and arrangement of these autumn colors, from some points of view, may not always result in the most perfect color-harmony. Indeed, the "loudness" of the Hudson River scenery in September has been the comment of more than one traveller in the United States; but this, I fancy, comes from considering the cubes of the mosaic separately instead of regarding the picture as a whole. Looked at in the part, the cold green of the pine may jangle with the scarlet of the maple, the blue of the sky may be out of key with the flaming sumacs along a bare hill-ridge; and in that way the autumn covering may be analyzed into something like a discord. But nature does not scatter its parts and leave them in any such helpless loneliness. There is just as much harmony in this pageant costume of the autumn as in the sombre grays and browns of the spring. For the same binding qualities of light and air are present. The autumn haze and the mellow glow from the sun cement all

The scarlet foliage.

Harmony of the scarlet landscape.

the parts together, blending them, toning them, binding them into a universal whole. Unity is, indeed, the key-note of all landscape; and it is the sweeping mass of this foliage, carpeting the hills and running over the meadows down to the scarlet reflection at the water's edge, that reveals and emphasizes the large harmony of the design.

Nature's sacrifices.

It is the mass and body of trees, too, that blend into unison the odd groupings wherein form, color, and texture are often recklessly sacrificed. Nature can and does throw away many effects that humanity would eagerly grasp. It rolls a whole mountain-side into one tone of green or yellow with scarcely a break, it ranks together acres of dark pines without a perceptible spot of white or yellow, it rears whole groves of white-trunked birches without a dark tree among them for relief or contrast. The landscape-gardener advises his client not to hang a weeping willow over a pool of water, but nature does it with impunity; the landscape-gardener advises contrasts of colors—yellow or light green against bottle-green; contrasts of texture—the fluffy leaf against the needle-point; contrasts of form—the short, stout tree against the tall, thin one; but nature seems to have

Tree-contrasts.

paid small attention to these canons of taste. It puts its growths together at random quite regardless of the part, but it is not so careless about the total result. The mass is always harmonious in its breadth.

The great volume of foliage undoubtedly has much to do with making the landscape in America harmonious, in spite of abrupt contrasts and vivid hues. The country is really exceptional in the extent of its timber-growths; and as for the rainbow foliage of September, one never sees elsewhere such a display. The vegetation of the tropics, which we vainly imagine corresponds to the brilliant plumage of a parrot or a bird of paradise, is on the contrary a mass of dark summer-green the year round; and many of the lands in the temperate zone show no great forest-color in the autumn. The foliage of the Northern United States and Canada has about it an incomparable richness, a vibrant sparkling quality which one cannot but think peculiar to the country itself. The traveller returning from Europe can feel a difference in the air and light as soon as he enters New York Harbor, and it is perhaps the air and the light that make possible the intense hues of foliage.

Tropical forests.

American forests.

European woodlands.

There are beautiful trees and groves in England, France, and Italy, and there are Black Forests, Bohemian Forests, and Harz Mountain woodlands in Germany, but the European woods are no improvement upon those of the Western continent. They are not so varied in form and color, nor have they the same freshness and wildness. In the Old-World forests it always seems borne in upon one that nature is playing the drudge to civilization, and that every large tree exists only to cast a protecting shadow over some house, park, or roadway. Happily, in this Western world, the flavor of the wilderness has not entirely departed. Along many a plain and valley nature rears its growths undisturbed, and upon many a mountain-side "the trees of the Lord" have not yet wholly become the trees of man.

CHAPTER XIV

EARTH COVERINGS

THE scientific distinction between a bush and a tree is simple, but somewhat arbitrary. It indicates a tree by its having a single stem or trunk, while the bush is peculiar in having several stems springing up from one root. But there is really no sharp division-line between the shrub and the larger growth. The one merges into the other. Regarded from a picturesque rather than a scientific point of view, there is a distinction just as arbitrary, which may be made after this fashion: The tree grows separately even in a forest, and its foliage begins so high up the trunk that the earth beneath it is usually exposed to view; the bush often grows in dense clumps over acres of ground, with foliage so close and so low that the earth is hidden from view. Perhaps then I may be allowed to treat of bushes under the general heading of "Earth Coverings," putting them in the same class with reeds and grasses.

Trees and shrubs.

Bush-growths.

The substitutes of nature.

Laurel and rhododendron.

The bushes make no such show on the face of the earth as the trees, though perhaps they cover more territory; and, moreover, they are frequently a secondary rather than a primary growth—a substitute rather than an original. Nature is fertile in resources, and wherever the earth is scarred by fire, tempest, or the axe, an effort is put forth to cover the spot with a new growth. Many of the shrubs and bushes and small-bunched thickets of the woods and hills are the result. In the coal regions of Pennsylvania, where the timber has been destroyed and many of the valleys have been turned into mere sluices and drainways for the black waters of coal mines, the laurel and the rhododendron grow in great profusion, covering valley, hill, and mountain for miles at a stretch. In the early summer, when they are in bloom, they are really splendid in effect. All the mountain seems in blossom, and along the ridges the color is banked up against the blue sky in pink and red clouds. In Southern California nature was probably never prodigal in the planting of forest-trees; but the neglect is atoned for by almost endless varieties of small bushes and trees that robe the mountains and the foothills in a mantle of many colors at all seasons of the year. Besides the evergreens there are the sumac, the white and lavender lilacs, the madroña, the manzanita, the wild mahogany, the choke-cherry, all rolled together along the hill-sides in great velvet waves fifteen feet or more in height. This is the chaparral—the dense thicket where the grizzly makes his home and breaks a path, where the mule-deer skulks at noonday, but where neither horse nor man finds easy thoroughfare. Desolate enough might be the hill-sides of California, were it not for this thick carpeting of bush and stunted tree. And were it not for the grease-wood, the sage-brush, and the spiny cactus, how very bare and dreary would be the alkaline plains! These growths of the arid lands are far from being joyous, but they are singularly appropriate to the landscape where they are seen. No other bushes, save these hardy shrubs, would live there, and nature does the best it can with every surface given it to care for.

California chaparral.

Sage-brush.

The clothing of the hills that lie along the Atlantic coast is something quite different from that of the Pacific slope or the plains. There is neither the density of the chaparral nor the meagreness of the sage-brush. The growth is more uniform. Oftentimes the laurel, the magnolia, the thorn, the dogwood, the hazel, tangled with sweet-brier, grape, and clumps of berry-bearing bushes may be seen in one landscape. There is no stint to the variety nor to the beauty of these growths, but simply because they spring up close beside us, and may be seen from almost any country dooryard, we are disposed to think them too common for admiration. Such a conclusion is of almost universal acceptance, but it is not the less shallow for that. It is the old error of thinking happiness in Rome or Athens or Bagdad rather than in our own heart and home. The unusual in nature is not by any means the most enjoyable. There is a greater charm in the commonplace, the humble things of the earth, if we have but the eyes to see them and the soul to feel them. A clump of hazel on the upland meadow, around which the daisies grow and through which the blackberry twines its white blossoms, may be a wonder-world of beauty if we study it in its form and color, its setting, light, and relation to the whole meadow. And the wild rose—the common wild rose—growing along the woodland road, unseen by the farmer's boy and the summer tourist, is a vision of loveliness beyond all description. How many

Upland bushes.

Common growths.

Wild roses.

times has it led poets to prove the poverty of language! With the dew upon it in the early morning, it is the fairest, purest growth in all the floral world. As children we knew it, plucked it, and scattered its petals upon the ground; but since then we have scarcely seen it. Grown to man's estate, we still walk along that woodland road on Sunday afternoons seeking a "breath of fresh air;" but we see little of anything. Our days of observation have passed and we have fallen upon days of reflection. Instead of looking without, our eyes are turned within, and we are studying some human problem, perhaps some business venture, while walking the new Garden of the Hesperides.

On the woodland road.

The bushes are the most varied in form and color of all the earth coverings, and they also form the densest shield against the sunlight. Sometimes, when they are scattered in broken clumps, the sunlit open spaces between them grow small grasses and weeds, but usually the hill-side bushes stand close together, the branches touching each other and throwing an almost perpetual summer shade upon the ground beneath. Naturally those plants only that live under shadow are found growing there. The moss clings to the rock, some thin grasses

Growths under shadow.

flourish, and occasionally, in the spring of the year, one meets with slender-stemmed wildflowers, looking pale and delicate in their shadowed homes. The hardy ferns will grow near the bush, but the ground they usually cover is under the forest-trees and in the oak openings. Everywhere, even in the Adirondack forests, their growth is rank. Sometimes they will reach up as high as one's head, but they are usually of knee-high growth and of a yellow-green hue. They do not usually grow well in the sunlight. Even the bracken of the Scotch hills and valleys clusters under the evergreen and the mountain-ash, or hides its roots beneath tall grass. It is a more rusty-looking covering than the American varieties of fern, but is nevertheless picturesque.

Fern and bracken.

The most conspicuous covering of Scotland, however, is the heather. It is a coarse, shaggy under-shrub, growing close to the soil and covering the treeless hills and moors in great fields many miles in extent. It belongs with the dark soil of the peat-beds, the crags of the mountain-peaks, and the low-flying clouds of Scotland, and is seen to advantage in the late summer when it is in bloom. The whole aspect of the country is then changed by it. One

Scotch heather.

hears it always spoken of as "purple heather," though in reality the coloring of the blossom is pink. Seen at a distance, however, especially at evening, it has a purplish effect which perhaps justifies the popular description of it. It grows in vast rolls, and sweeps along the slopes about Dalwhinnie in the Grampians, and nothing could be more beautiful than the hills of that region during the first week of September, when they are clad in their purplish-pink mantle. The absence of timber, the uniformity of the heather-covering, the beauty of the sky lines, the splendor of the light and the clouds, all make for a simple, yet broad and noble landscape—a country one might well fight for and, if need be, die for. *Heather-color.*

Our own golden-rod is no such complete earth covering as the heather, and it is not usually seen spread over such vast reaches of territory, but it nevertheless plays an important part in the autumn landscape. Oftentimes it covers many acres of field and upland, and in the mass of its coloring it is singularly rich and attractive. Very appropriate, too, is this coloring to the fall of the year when the skies are warming and the leaves are changing. In the late summer, when it first appears, it is a *Golden-rod.*

lemon-green, but as the flower opens into fuller bloom it changes to a clear, luminous chrome-yellow—a color that holds as a distinct hue for perhaps a greater distance than any other in nature's scale. Later on in the year, the golden-rod becomes faded and rusty, and is then contrasted with quantities of blue asters that grow up beside it and around it in the fields and meadows. In America it is in sort a national flower, growing tall and rank along almost every hill-side and roadway, and wherever growing lending mellowness and beauty to the landscape.

Blue asters.

The bushes, the ferns, the heather, and the golden-rod are coverings that belong distinctly to the uplands, the side-hills, and the mountain-slopes. The coverings that grow along the shore and upon the flat marshes and salt meadows are of an entirely different family. Some of them are grasses of thick, rank growth; others belong to the sedge group, and are even ranker in growth and darker in coloring than the grasses. The rush and the cat-tail grow along almost every coast and river delta where the ooze and mud washed down by streams give them a footing. I have already spoken of their great expansive beds and their varied

Rushes and flags.

coloring in summer and winter. A little farther back from the marsh, often quite close to it, are those dryer lands that grow tall grasses and weeds which are sometimes cut to make what is called "salt meadow hay." They do not make a strikingly beautiful growth, though they wave quite prettily in the wind, nor is the color of them in any way remarkable. Still farther back lie the pasture-lands and meadows where the ordinary field-grasses grow, and these are, perhaps, the most common of all the earth coverings.

Meadow growths.

There are some thirty-five hundred species of the grass family, ranging from the tall stalk of the bamboos to the small, almost moss-like buffalo grass of the plains. In the picturesque landscape they all have their place, not because they are different members of a botanical family and show slight variation in form and growth, but because they are all masses of fibre and color that carpet the open spots of the globe and lend to universal beauty. Nature did not, perhaps, grow them so much for picturesque effect as for use. They are the protectors of the soil from denudation by rains and frost. Wherever the surface of the earth is left bare, nature immediately starts the growth

The grasses.

of a covering, just as it heals an abrasion of skin on the human hand. The Indian trail, the bridle-path, even the track of the plough, are soon covered over and hidden by the creeping, weaving, intertwining grasses. The fields and meadows, where now the herbage grows thick and cattle graze, were perhaps but a few years ago sown with wheat and have only lately been allowed to "run to grass." The roots soon knit together and make a sod that rain does not wash and the stamp of many feet does not wear away.

Earth-protecting grasses.

And here in the meadow the grass grows rank, the buttercup spreads its yellow petals, the daisy and the dandelion flourish, and the wild violet springs up in little beds. Very commonplace is the ten-acre pasture, with its small knolls, its tufts of tall grass, its smooth-cropped interspaces, its wild-flowers, and its ivy-wound fence of stone; yet in this patched irregularity there is a whole world of loveliness. The quaint lines, the warmth and glow of color, and, above all, the broad area of sunlight, affect one emotionally. Take any man from the bustle of the city and place him there and he will instinctively breathe deeper, and though he may say little, yet be sure he is

Meadow and pasture.

making confession in his secret soul—confessing to a feeling he cannot define. A little swale of grass, a thistle, and a rock—what is there about them that cheapens the city street and the tall building? Is it anything more than that the one is natural and the other is artificial? We put blocks of stone together and try to create an impression of beauty such, perhaps, as nature produces; but the imitation falls far below the original. We rear spires and pinnacles in the air, palaces in the sun; but they are never so awe-inspiring as the mountains. We flatten the Fields of Mars; but they are not so impressive as the plains. We build baubles of form and color without number; but how petty they seem by comparison with nature's handiwork! A tree, a brook, or a hill—yes, even a flash of sunlight on a wayside flower—is worth them all. Honor to the work of man; honor to those who spin and carve and build; honor to the hand that rounded Peter's dome; but what of the Hand that rounded the earth and established the blue dome of the sky, what of the work of the Great Builder! *The natural vs. the artificial.*

And the wealth of color nature lavishes on the meadow and the pasture! With a prodigal hand she sacrifices half a dozen hues to *The wealth of colors.*

Meadow-flowers.

make one brilliant. The buttercup absorbs and practically annihilates green, red, blue, orange, violet; all these pass into the petals and are lost. Yellow alone it rejects and reflects, just as the violet throws back violet and the pink throws back pink. The white petal of the daisy, more imperious than the others, rejects all the hues and remains white or colorless; and there is a dark, bell-shaped wild-flower (its name I have never known) which absorbs all the hues and remains nearly black or colorless again. Yet with this enormous destruction of color that goes on, year in and year out the whole world round, nature never seems to want. To-day each woven thread of gold, silver, scarlet, or purple in her variegated garment throws off its light as brilliantly as in pre-Adamite days.

Pasture changes.

And how often the garment changes! Consider how many new robes the pasture-lot has in the course of the year—all of them bright and beautiful! There is the tender, yellow-green grass of early spring, which soon changes to dark green and is dotted with golden dandelions. When the dandelions have passed, the whole field turns yellow with buttercups, and is then blown white with daisies. In September it is silvered with wild oats, yellowed again with golden-rod or turned blue with asters. Finally, all is changed to russet and gray by frost, and at last buried under a white sheet of snow. The tall pine on the hill-top that has but one dress the whole year round—how much less care nature seems to have bestowed upon it than upon the pasture with its flowers! Yet we admire the pine, and perhaps care little for the pasture. We walk across the latter, treading the delicate grasses under foot, whipping off the heads of the daisies with our walking-stick, and thinking, perhaps, with Peter Bell that the meanest flower that blows is simply the meanest flower; but nature knows nothing of one creation meaner or nobler than another. It builds each thing perfect after its kind. Commonplace pasture and Olympian grove, mountain-crag, dense forest, gay flower, and lowly earth coverings are all of equal rank in nature's book of gold. Each has its measure of glory, each its peculiar beauty. *Nature's care.*

The cultivated grasses that cover the earth in spots, such as the fields of timothy and red clover, seem to have less charm than the wild growths, though no one can deny their beauty. The foaming whiteness of the blossoming buck- *Cultivated growths.*

wheat, the reddish hue of the ripened corn, the waving greens of the barley, oats, and wheat upon the hill-sides, are mere patches of local color, but they add greatly to the landscape; and where the bright yellow of ripe wheat is seen in vast masses, it is very impressive. The wheat-fields of Dakota and Minnesota, where once forty and fifty thousand acres of grain stood in unbroken reach from horizon to horizon, were almost as sublime as the ocean, and grander far in light and color than the tall grass of the prairies. Yet one can never escape the feeling that this is nature under the lash—nature more for man's sake than for her own sake. Her efforts are cramped to utility. The product is not what would be grown, but what must be grown. One cannot help feeling in the same way about the cultivated shrubs upon the lawn, and the flowers that grow in the Persian-carpet beds, the ugly little road-borders, and the glass houses. Beautiful they are, but their flush is hectic and they smell of the perfumery shop. They are nature's frailer children, and have not the vitality nor the wild, untamed beauty of the flowers growing on the meadows and the prairie.

And lastly, the smallest and the humblest of

the earth coverings—the mosses. Mountain, shore, plain, and meadow, each has its peculiar dress, and why not those spots of the dense woods where the straggling sunlight falls pale and broken on rocks and prostrate tree-trunks? The grasses and flowers will not grow there, save in isolated spots; the ground is too damp, the shade too dense. But these are the conditions of existence for those velvety growths with pin-like awns called the mosses. Flowerless, scentless, not brilliant in hue, and so humble in stature that we tread them under foot without seeing them, yet what a beautiful and perfect earth covering they make! Perhaps because they do not grow high they grow thick, forming a complete sod that rains and running waters cannot readily wash away. It is not a coarsely woven covering made up of many rough growths, but a compactly constructed mass. In these growths, which are placed where few see them, tucked away under rock bases, bunched about the roots of the great pines, or hidden under thick brush, it might be thought that nature would spare effort in perfecting the forms with nicety. But, no; every hair-root, every spore, every stem is wrought with a skill and a beauty

The mosses.

Moss structure.

Moss colors.

that would fit it to cover a royal throne. And the coloring of the mosses is not less wonderful. Chlorophyll is in their minute cells, as in those of the grasses and the leaves. The hue is green—evergreen in most of the wood-mosses—but what a variety in the color! You can hardly bring two pieces of moss together and find them of the same hue, because the conditions of light and moisture under which they grew were not the same. But none of the greens is harsh or discordant to the eye; from olive to green-gold all are harmonious, and all luxuriant in their depth of hue. Again, how soft and grateful to the touch the texture of the mosses! The awns that start up with the earliest awakening of spring are delicacy itself; and in the summer, when the tiny stems and leaves have woven their carpet of velvet, how pleasant it feels under the foot. The mosses were

Moss textures.

not designed to be walked upon by human feet, but, like the field-grasses, they are so constructed that human feet will not permanently injure them. Lying low on the ground, rain and hail fall upon them, snow covers them, frost binds them, but from none of these assaults comes harm. They were designed for places of exposure, but they were given a hardy, resistant nature and a compact surface to withstand the elements.

This is perhaps even truer of the gray lichens that cling to the loose bowlders on the mountain-side and color the barren crags and exposed rocks of the peak. They are the hardiest, and it is thought, geologically, among the earliest, of all plants, making a bed for the flora of the world by gathering about themselves grit and mould from the rock. Sun, wind, rain beat full upon them, but tenaciously they hold upon the stone, never moving, scarcely ever changing color. Sometimes called parasitic plants, they are really the protective coverings of the stone, as the mosses and the grasses are the coverings of the earth. The long-stemmed sea-weeds that cling about the coast bowlders—the *algæ* that ward off the thrust of waves and the grind of surges—are the ocean cousins of these mountain lichens. We know how the *algæ* cover and color the coast rocks, but we have, perhaps, less knowledge of the color-changes wrought on the mountain's peak by the lichens. The staining and what is called the "weather-beaten" look of rocks are largely their doing. The clean-faced bowlder dug from the soil and the lightning-broken surface on the mountain wall are

Gray lichens.

Rock-staining by lichens.

no sooner exposed to the light, the air, and the rain than they begin to darken and deepen in hue under the pencilling of the lichens. Even where the plant form is not recognizable, there is a grayish or greenish spot that tells of its coming. It may come slowly, for these hardy growths are never in a hurry to gain maturity. They know not time, yet they are never idle. Suns come and go and count out human years, but always with the lichens new spores are forming, new threads are creeping, new hues are gathering on stone and cliff and peak.

The work of mosses and lichens.

It seems a menial office—a humble part at best to play in this beautiful world—to protect and stain the rocks so that they shall withstand the elements and harmonize with the green of the trees and the blue of the sky; but how patiently the task is wrought, how faithfully the part is played! Green moss and gray lichen! The least pretentious of nature's creations they are, and yet how inevitably they again force a contrast with the handiwork of man! No human skill could weave such carpets; no dyes could produce such colors; no machines could stamp such patterns! The fabric is perfect of its kind.

Nature is above all! Unseen the loom, unseen the materials, unseen the hand of the weaver. Heat, light, and moisture—what simple ingredients, but brought together and working in unison what forms and colors they produce! what variety! what endless combination! And once again, as compared to the work of man, how permanent seems the product! Men and their deeds pass away, but nature seems immortal. The garden of the world is to-day as yesterday. Counting by human centuries we shall never know its decline. The light of the sun shall not be extinguished, and under it always the glow of the earth, the flame on the mountain-peak, the foam on the tossing wave. The blue sky, though it change its light from hour to hour, shall not diminish, and forever under its dome the drift of clouds, the fall of rain, the flash of the mountain-lake, and the glittering thread of the river winding downward to the sea. Shiftings of season and shiftings of color and foliage—change following change; but forever and forever the arrowy pine on the mountain, the golden-rod on the upland, and the flag by the sedgy shore.

And the great peace of it! Of what avail the struggle of races, the clashing of social systems, the ascending cry of the human! Serene above

The great peace.

it all, the Great Mother never hears, never heeds. The law of the kingdom—look to the law. Raise no hands of protest to the throne. There is no appeal. If you would cry out, go to the forest; if you would moan, stand on the prairie; if you would implore, look up at the sky and the sunlight. Learn from these. The law is written on them. Through all the ages of the earth's endurance that law has not failed to teach obedience, patience, peace.

www.ingramcontent.com/pod-product-compliance
Lightning Source LLC
Chambersburg PA
CBHW022043230426
43672CB00008B/1057